STOCHASTIC INTEGRATION

Michel Metivier

Ecole Polytechnique
Centre de Mathématiques Appliquées
Palaiseau, France

J. Pellaumail

Institut des Sciences Appliquées
Rennes, France

1980

ACADEMIC PRESS
A Subsidiary of Harcourt Brace Jovanovich, Publishers
New York London Toronto Sydney San Francisco

ACADEMIC PRESS, INC.
111 Fifth Avenue, New York, New York 10003

United Kingdom Edition published by
ACADEMIC PRESS, INC. (LONDON) LTD.
24/28 Oval Road, London NW1 7DX

Library of Congress Cataloging in Publication Data

Metivier, Michel.
Stochastic integration.

(Probability and mathematical statistics)
Bibliography: p.
Includes index.
1. Integrals, Stochastic. 2. Martingales (Mathematics) 3. Decomposition (Mathematics)
1. Pellaumail, Jean, joint author. II. Title.
QA274.22.M47 519.2 79-23096
ISBN 0-12-491450-0

PRINTED IN THE UNITED STATES OF AMERICA

80 81 82 83 9 8 7 6 5 4 3 2 1

CONTENTS

PREFACE

In writing this book we have attempted to make available to readers the experience that we have acquired in the past few years in dealing with stochastic integration.

This theory usually requires substantial prerequisites on the general theory of stochastic processes, especially if one wishes to consider discontinuous semimartingales, which occur more and more frequently in applications. We think the approach we have developed allows, on the contrary, a direct and rapid grasp on the subject, starting from basic knowledge of probability theory (probability spaces, independence, and conditioning).

The exposition here is therefore self-contained with respect to processes, stopping times, martingales, semimartingales, Brownian motion, etc. We have not aimed at completeness and emphasize what is basic in our opinion, which, in some instances, goes beyond what can be found in the available literature, and restrict ourselves to giving bibliographical references for subjects that are more naturally and well treated elsewhere.

For example, the development of stochastic partial differential equations and distributed systems has strenghtened the need for considering infinite-dimensional Hilbert-valued processes. We deal here with Hilbert-valued (sometimes, Banach-valued) processes. In many cases, this does not introduce really new difficulties, compared with the real case, at least as concerns initial results and properties. Actually, our experience is that methods that work in the same way for real and infinite-dimensional processes are often simpler than others designed only for the one-dimensional situation. When more sophisticated questions arise from the infinite dimensionality, this is mentioned and studied separately.

A main idea of the book is that the construction and properties of the stochastic integral need very little from the machinery of the general theory of stochastic processes. A difficulty arises when the question is how to characterize the class of processes that define a stochastic integral. It is now known that, in the real case, this class is exactly the class of semimartingales. We have also proved that the class of semimartingales possesses a majorization property, in this book called the $*$-domination property, which is crucial to simplifying the localization procedure and avoiding difficulties with unbounded jumps.

From these considerations, we have chosen the following exposition procedure.

In Chapter 1 we define in an elementary way and study the stochastic integral with respect to a class of processes, called π-processes, that satisfy some domination property. As a first step, we restrict ourselves to showing that this class contains sufficiently many processes to be interesting and is stable under change of variables (Ito formula, Chapter 2). The first two chapters thus contain as few general considerations on processes as possible and are entirely devoted to the stochastic calculus with π-processes. These π-processes will, in Chapter 4, turn out to be exactly the semimartingales in the real case.

In Chapter 3, after some classical elementary examples, we study (easily), using the $*$-domination property, a general stochastic integral equation, which contains all the stochastic integral equations considered by K. Ito, C. Doleans-Dade, N. Kamazaki, E. Protter, M. Emery, etc. We obtain existence, uniqueness, and stability properties for solutions and a nonexplosion criterion under conditions of Lipschitz type. We are interested only in what is usually referred to as strong solutions. There are specialized works on weak solutions for Ito equations, and we refer to them.

Chapter 4 is devoted to martingales and semimartingales. The materials presented here range from the basic equi-integrability properties of submartingales and stopping theorems to more sophisticated Burkholder-type inequalities. The Meyer decomposition theorem is proved in the real and Hilbert situations, and the relations between π-processes and semimartingales are elucidated. The reader will soon see that the adopted presentation is not quite traditional and contains new features which can sometimes be viewed as new results.

From Chapter 5 on we enter into more sophisticated theories, inasmuch as we need more advanced tools from functional analysis, with which the reader will be assumed to be rather familiar. The short Chapter 5 presents an aspect of stochastic integration that we have ourselves much explored in past years: connections between vector-valued measures and stochastic integrals. This could be a starting point for exposing stochastic integration. We preferred a more direct and elementary procedure in this book.

Chapter 6 deals with some special features of stochastic integration with respect to infinite-dimensional Hilbert space- (and sometimes Banach space-) valued martingales. It is important, in this case for example, to know how to integrate processes whose values may be discontinuous operators. The importance of stochastic disturbances, which in the modeling of distributed systems appear as generalized processes, led us to introduce a few basic notions on the stochastic integral with respect to cylindrical martingales and to construct in particular the so-called Radonifying integral, introduced first by B. Gaveau for infinite-dimensional Brownian motion.

In order to make the book as easy to use as possible, we have added a few extensions and examples after most sections and have included a subject index at the end.

The notation used in the book is listed on p. xi.

INTERDEPENDENCE TABLE

The different chapters of the book are partially independent. More precisely, the "logical" order relation among sections is the following (a dotted arrow A⇢B means that A is only partially useful for B).

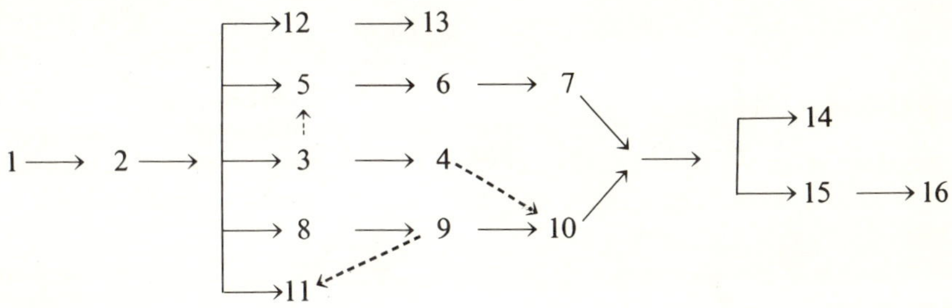

MEASURE-THEORETIC CONCEPTS

For measure-theoretic concepts, the reader is referred to any classical book on the subject in which real-valued random elements or real-valued processes are treated. Only a few of these books are mentioned in the bibliography ([Bau], [Nev-4], [Doo], for example).

When we speak of random variables X defined on a probability space $(\Omega, \mathfrak{F}, P)$ with values in a Banach space **B**, we always mean a mapping X from Ω into **B** such that the inverse image of the closed balls in **B** belongs to $\mathfrak{F}$. It is known from a Pettis theorem (see [Pet]) that when **B** is separable and $\mathfrak{F}$ is complete, this measurability property for X, often called *strong measurability*, is equivalent to the following: for every continuous linear form l on **B**, the real function $l \circ X$ is $\mathfrak{F}$-measurable.

For more detailed information, the reader is referred to [DuS], [Pet], [Bar], [Met-7].

ACKNOWLEDGMENTS

The authors wish to express their thanks to Professor E. Lukacs for having encouraged them to complete the writing of this book and to the many colleagues who made helpful remarks and suggestions.

They feel much indebted to the people at the Institut National des Sciences Appliquées and Ecole Polytechnique, who helped in typing the successive versions of the manuscript, especially to Mrs. Martinez and Mrs. Mouradian.

Finally, their thanks go to the staff at Academic Press for their meticulous labor and very cooperative attitude.

NOTATION

GENERAL MATHEMATICAL NOTATION

$\mathbf{R}$	set of real numbers
$\mathbf{R}_+$	set of positive or null real numbers
$\bar{\mathbf{R}}_+$	$\mathbf{R}_+ \cup \{+\infty\}$
$\mathbf{N}$	set of integers
$\bar{\mathbf{N}}$	$\mathbf{N} \cup \{+\infty\}$
x^+ (resp. x^-)	$\sup(x, 0)$ (resp. $\sup(-x, 0)$)
$x \wedge y$	infimum of two real numbers x and y
$x \vee y$	supremum of two real numbers x and y
$\mathcal{L}^0_{\mathbf{H}}(\Omega, \mathfrak{F}, P)$	set of $\mathbf{H}$-valued $\mathfrak{F}$-measurable functions on Ω, with the convergence in probability
$L^0_{\mathbf{H}}(\Omega, \mathfrak{F}, P)$	set of P-equivalence classes in $\mathcal{L}^0_{\mathbf{H}}(\Omega, \mathfrak{F}, P)$
$\mathcal{L}^p_{\mathbf{H}}(\Omega, \mathfrak{F}, P)$	vector space of p-integrable $\mathbf{H}$-valued $\mathfrak{F}$-measurable functions with the seminorm $\Vert f \Vert_p = [\int \Vert f \Vert^p \, dp]^{1/p}$ $(p \geqslant 1)$
$L^p_{\mathbf{H}}(\Omega, \mathfrak{F}, P)$ $(p \geqslant 1)$	Banach space of P-equivalence classes in $\mathcal{L}^p_{\mathbf{H}}(\Omega, \mathfrak{F}, P)$
$\mathfrak{F}(\Omega)$	set of subsets of Ω
1_A	indicator function of the set A
$A \backslash B$	difference of two sets A and B
$\mathfrak{B}_t$ (resp. $\mathfrak{B}_T$)	σ-algebra of Borel subsets of $[0, t]$ (resp. T)

NOTATION FOR VECTOR SPACES

All vector spaces considered are real vector spaces. When $\mathbf{B}$ is a Banach space, its dual (the set of continuous linear forms on $\mathbf{B}$) with the uniform norm topology is denoted $\mathbf{B}'$. The norm in $\mathbf{B}$ is written $\Vert \cdot \Vert_{\mathbf{B}}$ or simply $\Vert \cdot \Vert$ when no confusion is possible. The duality bilinear form between $\mathbf{B}$ and $\mathbf{B}'$ is denoted $\langle x, x' \rangle$.

For a Hilbert space $\mathbf{H}$ and two elements h, h' of $\mathbf{H}$, we usually write hh' for the scalar product of h and h'. When necessary, in order to avoid confusion, we write $\langle h, h' \rangle_{\mathbf{H}}$ or $(h \mid h')_{\mathbf{H}}$ for this scalar product.

$\mathcal{L}(\mathbf{B}; \mathbf{K})$ is the set of bounded linear operators from $\mathbf{B}$ into $\mathbf{K}$.

$\mathcal{L}_1(\mathbf{B}; \mathbf{K})$ is the set of nuclear operators from $\mathbf{B}$ into $\mathbf{K}$.

$\mathcal{L}_2(\mathbf{H}; \mathbf{G})$ is the set of Hilbert–Schmidt operators from the Hilbert space $\mathbf{H}$ into the Hilbert space $\mathbf{G}$.

The norm in $\mathcal{L}_1(\mathbf{H}; \mathbf{K})$ is written $\Vert \cdot \Vert_1$ and in $\mathcal{L}_2(\mathbf{H}; \mathbf{G})$, $\Vert \cdot \Vert_2$. The definitions of $\mathcal{L}_1(\mathbf{B}; \mathbf{K})$ and $\mathcal{L}_2(\mathbf{B}; \mathbf{K})$ are given in the text, Chapter 6, Section 14.

The tensor product spaces are also defined in the book: $\mathbf{H} \hat{\otimes}_2 \mathbf{H}$ in Section 3.4, $\mathbf{B} \hat{\otimes}_1 \mathbf{B}$ in Section 14.

NOTATION SPECIFIC TO THE BOOK

	Section
$\mathcal{A}$	1.7
$\mathcal{C}$	(processes) 9.6
d_X or $d(X)$	1.15
$\Delta X, \Delta M, \ldots$	4.6 or 6.3c
$\mathcal{E}, \mathcal{E}(\mathbf{L})$	2.2
$\mathcal{F}_t, \mathcal{F}_{t^+}, \mathcal{F}_{t^-}$	1.1
$\mathcal{F}_u$	1.2
$\mathcal{F}_u$	9.2
$L^2(X; \mathbf{L}; \mathbf{K})$	2.3
$L^*(M; \mathcal{P}; \mathbf{R}; \mathbf{K})$	14
$\Lambda^2(M; \mathcal{P}; \mathbf{B}; \mathbf{K})$	14
$\tilde{\Lambda}(\tilde{M}; \mathcal{P}, \mathbf{B}; \mathbf{K})$	16
$\mathcal{M}^2_T(\mathbf{H}), \mathcal{M}^2_T$	10.1
$\mathcal{O}$	1.21

	Section
$\mathcal{P}$	1.7
$\mathcal{R}$	1.7
Ω'	1
$[u], [u, v], \ldots$	1.3
X_{t^-}	1.4
X^u and $\bar{X}^u$	1.12, 1.13
$(\int Y\,dZ)$	2.4, 2.9
$[X]$	3.2
$\mathbf{[}X\mathbf{]}$	3.6
$[X, Y], \mathbf{[}X, Y\mathbf{]}$	4.1
$\langle X \rangle, \langle X, X \rangle$	10.2
$\boldsymbol{\langle} X \boldsymbol{\rangle}, \boldsymbol{\langle} X, X \boldsymbol{\rangle}$	10.2
$\boldsymbol{X^*}$	6.7

CHAPTER 1

STOCHASTIC INTEGRAL WITH RESPECT TO π-PROCESSES

This chapter is devoted to the construction and properties of the stochastic integral with respect to a very wide class of processes, called π-processes (primitive processes) here, which will later turn out to contain (and in the real case actually be equal to) the class of so-called semimartingales as defined by P. A. Meyer (cf. [Mey-3 or 4]).

Since we intend to keep the book as self-contained and elementary as possible, we concentrate in Section 1 all the basic notions and properties related to stochastic processes, necessary for our further development. The more elaborate results concerning, for example, the Meyer-decomposition theorem and semimartingales are not given until they are needed in Chapter 4. In Section 2 we define L^2-π-processes and π-processes before giving the construction of the stochastic integral.

1 STOCHASTIC BASIS AND PROCESSES

STOCHASTIC BASIS AND STOPPING TIMES

1.1 Stochastic Basis: Definition

Let T be a subset of the real line. We call a *stochastic basis* a family $(\Omega, \mathcal{F}, (\mathcal{F}_t)_{t\in T})$ such that $(\Omega, \mathcal{F})$ is a measurable space and $(\mathcal{F}_t)_{t\in T}$ is an increasing family of sub-σ-algebras of $\mathcal{F}$ called a *filtration*. If $(\Omega, \mathcal{F}, P)$ is a probability space, we call the family $(\Omega, \mathcal{F}, P, (\mathcal{F}_t)_{t\in T})$ a *probabilized stochastic basis*.

In all parts of this work except for Chapter 5, T is an interval of $\overline{\mathbf{R}}^+$ or $\overline{\mathbf{N}}$, where $\mathbf{R}$ is the real line and $\mathbf{N}$ the set of integers. Moreover, we always assume that 0 belongs to T. More precisely, we usually have $T = \mathbf{N}$ or $T = [0, t_m]$, $t_m < \infty$ or $T = \mathbf{R}$. When we do not specify whether or not T is bounded, we write t_∞ for the supremum of T in $\overline{\mathbf{R}}^+$ and we define $T' := T\backslash\{0\}$ and $\Omega' := \Omega \times T'$.

Throughout Section 1, we consider a probabilized stochastic basis $(\Omega, \mathcal{F}, P, (\mathcal{F}_t)_{t\in T})$. We say that this basis is *complete* if the space $(\Omega, \mathcal{F}, P)$ is a complete probability space and if for every element A of $\mathcal{F}$ such that $P(A) = 0$ and for every element t of T, A belongs to $\mathcal{F}_t$.

For each element t of T we define $\mathcal{F}_{t^+} := \bigcap_{\epsilon>0} \mathcal{F}_{t+\epsilon}$, and we say that the family $(\mathcal{F}_t)_{t\in T}$ is *right continuous* if $\mathcal{F}_t = \mathcal{F}_{t^+}$ for each element t of T (for every $t \in T$, $\mathcal{F}_{t^-}$ denotes the σ-algebra generated by $\bigcup_{s<t,\, s\in T} \mathcal{F}_s$).

If $\mathbf{H}$ is a Banach space (with its σ-algebra $\mathcal{H}$ of Borel sets), we write $L^0_{\mathbf{H}}(\Omega, \mathcal{F}_t, P)$ for the complete metric space obtained by endowing the set of $\mathbf{H}$-valued $\mathcal{F}_t$-measurable random variables with the metric of convergence in probability.

1.2 Stopping Times and Associated σ-Algebras

Let u be a measurable mapping from $(\Omega, \mathcal{F})$ into $(T \cup \{\infty\}, \mathcal{B})$ where $\mathcal{B}$ is the σ-algebra of Borel subsets of $T \cup \{\infty\}$. The mapping u is called a stopping time if for every $t \in T$ the set $\{\omega : u(\omega) \leqslant t\}$ belongs to $\mathcal{F}_t$.

If u and v are two stopping times, it can easily be seen that $u \vee v$ and $u \wedge v$ are also stopping times.

If u is a stopping time, we denote by $\mathcal{F}_u$ the σ-algebra defined by

$$\mathcal{F}_u := \{A : A \in \mathcal{F} \text{ and } \forall t \in T, (A \cap [u \leqslant t]) \in \mathcal{F}_t\}.$$

If s belongs to T and if $u = s$ (for every ω), we see that $\mathcal{F}_u = \mathcal{F}_s$. There is thus no confusion of notation possible when we write s for the constant stopping time equal to $s \in T$.

1.3 Stochastic Intervals

Let u and v be two stopping times. We shall denote the following subset of $\Omega \times T$ by $]u, v]$:

$$]u, v] = \{(\omega, t) : u(\omega) < t \leqslant v(\omega)\} \cap (\Omega \times T).$$

(Notice that if $v(\omega) = \infty$, $(\omega, \infty) \notin]u, v]$ when $\infty \notin T$.)

The sets $[u, v], [u, v[, \ldots$ are defined in a clear analogous way. All such sets $[u, v[,]u, v], \ldots$ are called *stochastic intervals.*

If u and v are elements of T and if we still denote by u and v the constant stopping times with values u and v, respectively, the notation $]u, v]$ is ambiguous inasmuch as it stands for a subset of T and a subset of $\Omega \times T$ as well. But usually the meaning is clear from the context. We write $[u]$ for the *graph* $[u, u]$ of u.

PROCESSES

1.4 Processes (Definitions) and Cadlag Notation

The word "process" has several different meanings in probability theory.

If $(\mathbf{H}, \mathcal{H})$ is a measurable space, we shall say that X is an $\mathbf{H}$-valued process if X is an $\mathbf{H}$-valued mapping defined on $(\Omega \times T)$ and X_t is $\mathcal{F}$-measurable for every $t \in T$.

We shall say that X is a "process *defined up to modification*" if $X = (X_t)_{t \in T}$ is a mapping from T into $L^0_{\mathbf{H}}(\Omega, \mathcal{F}, P)$.

If X and X' are two processes, X' is said to be a *modification* of X if for each element t of T, $X_t = X'_t$ a.s.

Let $\mathbf{H}$ be a topological space. Let f be an $\mathbf{H}$-valued function defined on T; it is called cadlag if it is right continuous and has a left limit in every $t \in T$ (in French, "f est continue à droite et admet une limite à gauche").

An $\mathbf{H}$-valued process X is said to be *cadlag* if, for every element ω of Ω, the *sample function* $t \rightsquigarrow f(t) = X_t(\omega)$ is cadlag as defined above.

The sample function $t \rightsquigarrow X_t(\omega)$ is also called a *path* of X.

We shall also use the notations *caglad* (left continuous with a right limit) and *laglad* (with left and right limits).

In the same way, we call a process X continuous (resp. right continuous) if, for every element ω of Ω, the sample function $t \rightsquigarrow X_t(\omega)$ is continuous (resp. right continuous).

Two processes X and X' are *P-equivalent* iff $P(\{\omega : \exists t, X_t(\omega) \neq X'_t(\omega)\}) = 0$.

A process with values in a vector space is often called *evanescent*, if it is P-equivalent to the null process (i.e., the process whose path is identically zero).

Most of the processes considered in the sequel will be actually defined up to P-equivalence. In this case, a process X will be said to be continuous, cadlag, etc. if it is P-equivalent to a continuous, cadlag, etc. process.

If X is a cadlag process, we denote by $(X_{t^-})_{t \in T}$ the caglad process $X_{t^-}(\omega) := \lim_{s \uparrow t} X_s(\omega)$ for every $t \in T$ $(t \neq 0)$ and $\omega \in \Omega$, and $X_{0^-} := X_0$.

Two cadlag processes X and Y such that X is a modification of Y are clearly P-equivalent: we have indeed $X_t(\omega) = Y_t(\omega)$ for all $t \in T$ except when there is a rational number q such that $X_q(\omega) \neq Y_q(\omega)$, and such a possibility occurs only for a P-null set of Ω when X is a modification of Y.

A process X or a process X defined up to a modification is said to be *adapted* (with respect to the stochastic basis $(\Omega, \mathcal{F}, P, (\mathcal{F}_t)_{t \in T})$) if for every $t \in T$ the random variable $X_t(\cdot)$ belongs to $\mathcal{L}^0_{\mathbf{H}}(\Omega, \mathcal{F}_t, P)$.

Let u be a $T \cup \{\infty\}$-valued function defined on Ω and measurable with respect to $\mathcal{F}$ and $\mathcal{C}$. It can easily be seen that the definitions imply that u is a stopping time if and only if the process $X = 1_{[0, u[}$ is adapted.

EXAMPLES

1.5 Examples: Brownian Motion and Poisson Processes

Brownian motion An $\mathbf{R}^n$-valued process X on an interval T is called a brownian motion adapted to the filtration $(\mathcal{F}_t)_{t \in T}$ if it possesses the following properties:

[I] for every s and $t \in T$ with $s < t$, the random variable $X_t - X_s$ is $\mathcal{F}_t$-measurable and independent of the σ-algebra $\mathcal{F}_s$ for the probability P;

[B] for every s and $t \in T$ with $s < t$, the random variable $X_t - X_s$ is Gaussian with mean 0 and covariance matrix $\sigma^2(t-s)I_n$, where σ^2 is a given scalar and I_n the $n \times n$ identity matrix.

The process is said to start at $x \in \mathbf{R}^n$ if $X_0 = x$ a.s. It can be directly proved (see [ItM] or [Nev-4]) that every brownian motion has a modification with continuous paths. This is also a consequence of a more general theorem (see Section 1.17). From now on, we shall always consider continuous versions of brownian motion.

A canonical brownian motion is a brownian motion $(\Omega, \mathcal{F}, (\mathcal{F}_t)_{t \in T}, (X_t)_{t \in T}, P)$, where Ω is the set $C(T, \mathbf{R}^n)$ of continuous functions from T into $\mathbf{R}^n$, X_t the projection mapping $\omega \rightarrow X_t(\omega) := \omega(t)$ for every $t \in \mathbf{R}^+$, $\mathcal{F}_t$ the σ-algebra generated by $\{X_s : s \leqslant t\}$, and $\mathcal{F}$ the σ-algebra generated by $\bigcup_{t \in T} \mathcal{F}_t$.

A canonical brownian motion is uniquely determined by σ and the starting point $x \in \mathbf{R}^n$. The measure P corresponding to $\sigma = 1$ and $x = 0$ is often called the Wiener measure on $C(T, \mathbf{R}^n)$.

For more information on brownian processes, see [ItM].

Poisson process An $\mathbf{N}$-valued process X on an interval T is called a Poisson process adapted to the filtration $(\mathcal{F}_t)_{t \in T}$ if it possesses the property [I] above and

[P] for every s and $t \in T$ with $s < t$, the integer-valued random variable $X_t - X_s$ obeys a Poisson law with parameter $\lambda(t-s)$, where λ is a given positive constant

$$P(\{X_t - X_s = n\}) = \frac{\lambda^n (t-s)^n}{n!} e^{-\lambda(t-s)}.$$

With probability one, the paths of a Poisson process have only finitely many jumps on every finite time interval, these jumps being of magnitude one. Every Poisson process has therefore a modification that is cadlag. In the sequel, only these cadlag versions of Poisson processes will be considered.

1.6 An Example of Stopping Time

Lemma *Let* **H** *be a Banach space. Let X be an* **H**-*valued adapted process, right or left continuous. Let u be a stopping time and a be a real number. For each element ω of Ω, we define*

$$v(\omega) := \inf\{t : t > u(\omega), \|X_t(\omega) - X_{u(\omega)}(\omega)\| > a, t \in T\}$$

if $u(\omega) \in T$ and the above set is not empty, and $v(\omega) := \infty$ otherwise. Then v is a stopping time with respect to the family $(\mathcal{F}_{t^+})_{t \in T}$.

Proof Let Q' be any denumerable dense set in T with $t_\infty \in Q'$ if $t_\infty < \infty$, and let $(S(n))_{n>0}$ be a sequence of finite subsets of Q' increasing to Q'. We set

$$v'(\omega) := \inf\{t : t \in Q', t \geqslant u(\omega), \|X_t(\omega) - X_{u(\omega)}(\omega)\| > a\},$$

$$v_n(\omega) := \inf\{t : t \in S(n), t \geqslant u(\omega), \|X_t(\omega) - X_{u(\omega)}(\omega)\| > a\}$$

(with the convention $v'(\omega) := \infty$ or $v_n(\omega) := \infty$ if the above sets are empty).

It can easily be seen that for each element ω of Ω, $v'(\omega) = v(\omega)$ and $v'(\omega) = \inf_{n>0} v_n(\omega)$. Since the mapping $(\omega, s) \rightsquigarrow X_s(\omega)$ restricted to $\Omega \times [0, t]$ is easily seen to be $\mathcal{F}_t \otimes \mathcal{B}_{[0,t]}$-measurable, the set

$$\{v_n \leqslant t\} = \bigcup_{\substack{s \in S_n \\ s \leqslant t}} \{\omega : \|X_s(\omega) - X_{u(\omega)}(\omega)\| > a\} \cap \{t \geqslant u\}$$

$$\text{for every} \quad t \in S(n),$$

belongs to $\mathcal{F}_{t^+}$. Therefore v_n is a stopping time for every n. Then, we have only to prove that the limit v of a decreasing sequence $(v(n))_{n>0}$ of stopping times is a stopping time for the family $(\mathcal{F}_{t^+})_{t \in T}$.

Let t be an element of T; we set

$$A := \{\omega : v(\omega) > t\} \qquad \text{and} \qquad A(n, k) := \{\omega : v_n(\omega) > t + 1/k\}.$$

We have $A = \bigcup_{k>0}\{\bigcap_{n>0} A(n, k)\}$. Moreover, the set $\bigcap_{n>0} A(n, k)$ belongs to $\mathcal{F}_{t+1/k}$; thus A belongs to $\mathcal{F}_{t+1/k}$ for every integer k and A belongs to $\mathcal{F}_{t^+}$. This proves that v is a stopping time with respect to the family $(\mathcal{F}_{t^+})_{t \in T}$.

PREDICTABLE SETS AND PROCESSES

1.7 Predictable Sets; Notations $\mathcal{R}$, $\mathcal{A}$, and $\mathcal{P}$

We call $\mathcal{R}$ the family of subsets A of $\Omega' = \Omega \times T'$ of the form $A = F \times]s, t]$, where F belongs to $\mathcal{F}_s$ and $s, t \in T$. We denote by $\mathcal{A}$ the boolean ring generated by $\mathcal{R}$ on Ω', by $\mathcal{P}$ the σ-algebra generated by $\mathcal{R}$ (or $\mathcal{A}$): the elements of this σ-algebra are called the predictable sets. An $(H, \mathcal{H})$-valued process is said to be predictable if, restricted to Ω', this process is measurable with respect to $\mathcal{P}$ and $\mathcal{H}$. If T is an interval of the form $[0, t_\infty]$, then $\mathcal{A}$ is clearly an algebra.

1.8 Decomposition of Each Element of $\mathcal{A}$

Lemma *If A is an element of $\mathcal{A}$, there exists a finite family $(A_i)_{i \in I}$ of elements of $\mathcal{R}$, which is a partition of A.*

Proof Let $\mathcal{A}'$ be the class of all elements A of $\mathcal{A}$ for which there exists a finite family $(A_i)_{i \in I}$ of elements of $\mathcal{R}$ which is a partition of A. To prove $\mathcal{A} = \mathcal{A}'$ it is sufficient to prove that $\mathcal{A}'$ is a boolean ring. For that it is sufficient to prove that if A and B are elements of $\mathcal{A}$, the same holds for $A \setminus B$. Then, suppose $B = \bigcup_{i=1}^n B_i$, where $(B_i)_{1 \leqslant i \leqslant n}$ is a finite family of elements of $\mathcal{R}$. We define C_i recursively by $C_1 = A$ and $C_{i+1} = C_i \setminus B_i$. We have $C_{n+1} = A \setminus B$. Reasoning by recurrence, we are led to prove that if D is an element of $\mathcal{A}'$, it is the same for $D \setminus B_i$; but the proof reduces to considering D in $\mathcal{R}$, and in this case, the property is easy to check.

1.9 $\mathcal{A}$ and the Stochastic Intervals $]u, v]$

Lemma *The ring $\mathcal{A}$ is identical with the ring $\mathcal{A}'$ generated by the stochastic intervals $]u, v]$ where u and v are simple T-valued stopping times (i.e., such that the sets $u(\Omega)$ and $v(\Omega)$ are finite).*

Proof (1) First, we prove that $\mathcal{A} \subset \mathcal{A}'$. It is sufficient to show that if $B := F \times]s, t]$ is an element of $\mathcal{R}$, then B is also an element of $\mathcal{A}'$; but $B =]u, v]$, where u and v are stopping times defined by $v(\omega) := t$, $\forall \omega$, and $u(\omega) := t$ if $\omega \in (\Omega \setminus F)$ and $u(\omega) := s$ if $\omega \in F$.

(2) We now prove that $\mathcal{A}' \subset \mathcal{A}$. Let u be a simple finite stopping time. Then there exists a finite increasing sequence $(t(k))_{1 \leqslant k \leqslant n}$ of elements of T and an associated sequence $(F(k))_{1 \leqslant k \leqslant n}$ of elements of $\mathcal{F}$ such that

(a) for every integer k, $F(k)$ belongs to $\mathcal{F}_{t(k)}$,

(b) $(F(k))_{1 \leqslant k \leqslant n}$ is a partition of Ω, and

(c) $u = \sum_{k=1}^{n} t(k) 1_{F(k)}$.

Then if for every k we set $B_k := (F(k) \times]t(k), a])$, where $a \in T$ and $a > u$, $(B(k))_{1 \leqslant k \leqslant n}$ is a partition of $]u, a]$. This shows that $]u, a]$ is an element of $\mathcal{Q}$. Since $]u, v] =]u, a] \setminus]v, a]$ for any $a \geqslant v$, the result is thus proved.

1.10 Predictable Sets with Respect to the Family $(\mathcal{F}_{t^+})$

Proposition *The σ-algebra $\mathcal{P}$ of predictable sets associated with the family $(\mathcal{F}_t)_{t \in T}$ is the same as the σ-algebra $\mathcal{P}^+$ of predictable sets associated with the family $(\mathcal{F}_{t^+})_{t \in T}$.*

Proof We have $\mathcal{P}^+ \supset \mathcal{P}$ from the definition. Moreover, if H is an element of $\mathcal{F}_{s^+}$, we have $H \times]s, t] = \bigcup_{k>0} (H \times]s + (1/k), t])$ where, for every integer k, $H \times]s + (1/k), t]$ is an element of $\mathcal{P}$. Then $\mathcal{P}^+ \subset \mathcal{P}$.

1.11 Left-Continuous Processes and Predictable Processes

Proposition *Let us assume $T = \mathbf{R}_+$. For every stopping time (resp. bounded stopping time) u, there exists a sequence $(u(n))_{n>0}$ of simple (resp. simple bounded) stopping times with $u(n) > u$ on the set $\{u < \infty\}$ which decreases to u; thus the real process $1_{]0, u]}$ is a predictable process. More generally, a Banach space valued left continuous adapted process is a predictable process.*

Proof (1) We define

$$u(n) := \begin{cases} \sum_{0 \leqslant k < n2^n} (k+1) 2^{-n} 1_{\{k2^{-n} \leqslant u < (k+1)2^{-n}\}} & \text{on } \{u < n\} \\ +\infty & \text{on } \{u \geqslant n\}. \end{cases}$$

It is immediately verified that the $u(n)$ are stopping times, which are bounded as soon as u is bounded by n.

The fact that $(u(n))_{n \in \mathbf{N}}$ is a decreasing sequence converging to u with $u(n) > u$ is trivial from the definition.

(2) Since $]0, u] = \bigcup_{p \geqslant 0} \bigcap_{n \geqslant p}]0, u(n) \wedge p]$ and $]0, u(n) \wedge p] \in \mathcal{Q}$, the set $]0, u]$ is clearly predictable.

(3) More generally, let X be a Banach-valued left continuous adapted process.

We define

$$X^n(\omega, t) := \sum_{k \leqslant 0} X_{k2^{-n}}(\omega) 1_{]k2^{-n}, (k+1)2^{-n}]}(t) + X_0(\omega) 1_{\{0\}}(t).$$

Since every random variable $X_{k2^{-n}}$ can be approximated for pointwise convergence on Ω by simple $\mathcal{F}_{k2^{-n}}$-measurable random variables, the processes (see (2) above) $X_{k2^{-n}}1_{]k2^{-n},(k+1)2^{-n}]}$ are predictable. The predictability holds therefore for X^n. The left continuity of X implies the convergence of $X^n(\omega, t)$ to $X(\omega, t)$ for all ω and t. From this follows the predictability of X.

LOCALIZATION AND PRELOCALIZATION

1.12 Stopped Process and Localization

Definition Let u be a T-valued random variable defined on $(\Omega, \mathcal{F})$ and X a process. Let $\overline{X}^u$ be the process defined by

$$\overline{X}_t^u(\omega) = X_t(\omega) \qquad \text{if} \quad t \leqslant u(\omega),$$

$$\overline{X}_t^u(\omega) = X_{u(\omega)}(\omega) \quad \text{if} \quad t \geqslant u(\omega).$$

$\overline{X}^u$ is called the process stopped at the random time u.

If u is a stopping time and if X is a right continuous adapted process, it can easily be seen that $\overline{X}^u$ is also an adapted process.

Let X be a process on $T \subset \mathbf{R}^+$. It is often useful to consider an increasing sequence $(u(n))_{n>0}$ of stopping times such that $\lim_n P\{u(n) < t\} = 0 \ \forall t \in T$, and to consider the processes $\overline{X}^{u(n)}$ which are the process X stopped at the stopping times $u(n)$. This procedure is called *localization*. In this situation, one says that X is *locally bounded*, *locally measurable*, etc. if each process $\overline{X}^{u(n)}$ (which is the process X stopped at the stopping time $u(n)$) is bounded, measurable, etc.

1.13 Prelocalization

Let u be a T-valued stopping time and X be a cadlag process; let X^u be the process defined by

$$X_t^u(\omega) = X_t(\omega) \qquad \text{if} \quad t < u(\omega),$$

$$X_t^u(\omega) = X_{u(\omega)-}(\omega) \quad \text{if} \quad t \geqslant u(\omega).$$

We shall say that X^u is the process X *stopped strictly before the stopping time* u. If X is adapted, it is the same for X^u. As in Section 1.12, it is often convenient to consider a sequence $(u(n))_{n>0}$ of stopping times and the sequence $(X^{u(n)})_{n>0}$ of associated processes. This procedure will be called *prelocalization* when $\forall t \in T$, $\lim_{n\to\infty} P\{u(n) < t\} = 0$. If for each integer n, $X^{u(n)}$ is bounded, continuous, etc., X will be said to be *prelocally bounded*, *prelocally continuous*, etc.

1.14 Predictable Stopping Times

Definition Let u be a stopping time. We call u predictable if there exists a sequence $(u(n))_{n>0}$ of stopping times increasing to u and such that for each integer n and each element ω of Ω, with $u(\omega) > 0$:

$$[u(n)](\omega) < u(\omega).$$

In this case $]0, u[= \bigcup_{n>0}]0, u(n)]$ is a predictable set.

As a consequence of the proposition in Section 1.11, the set $[u] \cap \Omega'$ is predictable. There is a converse property (see Section 8, Complements).

For an example of a nonpredictable set, see Exercise 2 in Section 1.23.

DOLÉANS FUNCTION OF A PROCESS. EXISTENCE OF CADLAG MODIFICATIONS

1.15 Doléans Function

Definition Let X be a process or a process defined up to modification with values in the Banach space H such that for every t in T, X_t belongs to $L^1_{\mathsf{H}}(\Omega, \mathfrak{F}, P)$. For each element $A = F \times]s, t]$ of $\mathfrak{R}$, we set

$$x(A) := E[1_F(X_t - X_s)].$$

It can easily be seen that x can be extended in a unique way into an additive function defined on $\mathfrak{A}$.

We shall denote this function by $d(X)$ and call it the *Doléans function* of the process X.

We are mainly interested in the case in which $d(X)$ is σ-additive: in this case, $d(X)$ is called the Doléans measure of the process X (cf. [Dol-3]).

Later (Section 8), sufficient conditions for the σ-additivity of $d(X)$ will be given (see Section 1.20 for a particular case).

1.16 Existence of Cadlag Modifications for Processes with Bounded $d(X)$

For a process X with values in $\mathbf{H}$ and for every predictable rectangle $A := F \times]s, t]$, we denote by $X(A)$ the random variable

$$X(A) := 1_F(X_t - X_s).$$

It is clear that $A \rightsquigarrow X(A)$ extends into an additive mapping from $\mathfrak{A}$ into the set of $\mathbf{H}$-valued random variables. If $E\|X_t\| < \infty$ for all t, then $E(X(A)) = d(X)(A)$ as previously defined.

We recall that $L^0_{\mathbf{H}}(\Omega, \mathfrak{F}, P)$ is the vector space of $\mathbf{H}$-valued random variables endowed with the topology of convergence in probability (see Section 1.1).

Theorem *Let X be an adapted process defined up to modification on an interval $T \subset \overline{\mathbf{R}}^+$, with values in a finite-dimensional vector space $\mathbf{H}$ and right continuous in probability (i.e., for every $s \in T$ and every $\epsilon > 0$: $\lim_{t \downarrow s} P[\|X_t - X_s\| > \epsilon] = 0$). We suppose that X satisfies one of the following two properties:*

(i) *for each element t of T, X_t is an element of $L^1_{\mathbf{H}}(\Omega, \mathfrak{F}, P)$ and the set $\{z : z = [d(X)](A), A \in \mathcal{Q}\}$ is bounded in $\mathbf{H}$;*

(ii) *the set $\{z : z = X(A), A \in \mathcal{Q}\}$ is bounded in $L^0_{\mathbf{H}}(\Omega, \mathfrak{F}, P)$.*

Then there exists a cadlag process Y, defined up to P-equivalence, which is a modification of X.

Proof It is sufficient to consider the case in which $T = [0, t_m]$, $t_m < \infty$; $t_m \in \mathbf{Q}$. It is also sufficient to consider the case in which X is a real process, by considering coordinates in $\mathbf{H}$.

Condition (ii) can be formulated as follows:

(ii′) there exists a positive decreasing function f defined on $\mathbf{R}^+$ such that $\lim_{x \to \infty} f(x) = 0$ and for every A in $\mathcal{Q}$ and every strictly positive number d we have $P\{|X(A)| > d\} \leqslant f(d)$.

Let Q' be the set of the rational numbers belonging to T. For every t in Q', we put $Z_t := X_t$.

(1) We first prove that the process $(Z_t)_{t \in Q'}$ has almost surely paths with left and right limits at every $t \in Q'$. Let (a, b) be a pair of rational numbers with $a < b$. Let S be a finite part of Q' and $\{t(k)\}_{1 \leqslant k \leqslant n}$ be the increasing ordered family of the elements of S. Let $\{u(k)\}_{1 \leqslant k \leqslant 2n}$ be the associated family of simple stopping times defined by recurrence by

$$\begin{aligned} u(1) &:= 0; \\ u(2k+1) &:= \inf\{t : t \in S, t \geqslant u(2k), Z_t > b\} \wedge t_m, \\ u(2k) &:= \inf\{t : t \in S, t \geqslant u(2k-1), Z_t < a\} \wedge t_m. \end{aligned}$$

Let $A(j, s) \subset \Omega$ be the domain where the process $(Z_t)_{t \in S}$ has more than $(j-1)$ upcrossings of the interval $[a, b]$; if $\omega \in \Omega$, either $\omega \in A(j, S)$, and this implies

$$\sum_{k=1}^{j} \left[Z_{u(2k+1)} - Z_{u(2k)} \right] \geqslant j(b-a),$$

or $\omega \notin A(j, S)$, and this implies

$$\sum_{k=1}^{j} \left[Z_{u(2k+1)} - Z_{u(2k)} \right] \geqslant -(Z_{t_m} - a)^-$$

(we have $Z_{u(2k+1)} - Z_{u(2k)} < 0$ only if $Z_{u(2k)} < a$ and $Z_{u(2k+1)} = X_{t_m}$). Then,

$$\sum_{k=1}^{j} \left[Z_{u(2k+1)} - Z_{u(2k)} \right] \geqslant j(b-a)1_{A(j,S)} - (Z_{t_m} - a)^- 1_{\Omega \setminus A(j,S)}.$$

If condition (i) is fulfilled, we put

$$C_j := \frac{1}{j(b-a)} \left\{ E(|Z_{t_m} - a|) + \sup_{A \in \mathscr{Q}} |[d(X)](A)| \right\}.$$

If condition (ii) (i.e., (ii′)) is fulfilled, we put

$$C_j := f[(j/2)(b-a)].$$

In any case we have

$$P[A(j,S)] \leqslant C_j \qquad \text{and} \qquad \lim_{j \to \infty} C_j = 0.$$

Now, we consider an increasing sequence $(S(n))_{n>0}$ of finite parts of Q' such that $Q' = \bigcup_{n>0} S(n)$. Let $A(j, Q')$ be the domain where the process $(Z_t)_{t \in Q'}$ has more than $(j-1)$ upcrossings of the interval $[a, b]$; we have $A(j, S(n)) \uparrow A(j, Q')$, and this implies

$$P[A(j, Q')] \leqslant C_j.$$

Thus if $A(Q')$ is the domain where the process $(Z_t)_{t \in Q'}$ has an infinity of upcrossings of the interval $[a, b]$, we have

$$P[A(Q')] = 0.$$

Considering the family of all pairs of rational numbers a and b with $a < b$, we see that with probability one there do not exist two different real numbers c and d such that Z upcrosses the interval $[c, d]$ infinitely often. This is clearly equivalent to saying that the process Z is a.s. ladlag (i.e., with left and right limits; see Section 1.4).

(2) Now for almost all $\omega \in \Omega$ and for every $t \in T$ we may define

$$Y_t(\omega) := Z_{t^+}(\omega) := \lim_{s \downarrow t} Z_s(\omega).$$

Let t be an element of T and $\{t(k)\}_{k>0}$ be a sequence of elements of Q' decreasing to t; the sequence of random variables $(Y_{t(k)})_{k>0}$ converges a.s. to Y_t (by the definition of Y_t) and in probability to X_t; hence $Y_t = X_t$ a.s. and Y is a modification of X.

Remark It is proved in [Met 6] that the analogous statement holds with property (i) for processes with values in a Banach space having the so-called Radon–Nikodym property. This is therefore true for reflexive Banach spaces and, in particular, for Hilbert spaces. However we do not want to enter into these considerations here.

MARTINGALES AND SUBMARTINGALES: BASIC PROPERTIES

1.17 Martingales

Definition Let X be a process or a process defined up to modification, with values in the Banach space $\mathbf{H}$ such that for every t in T, X_t belongs to $L^1_{\mathbf{H}}(\Omega, \mathcal{F}_t, P)$. We say that X is a *martingale* if the associated Doléans function is identically null.

It can easily be seen that this condition is equivalent to saying that for each pair (s, t) of elements of T with $s < t$ we have

$$E(X_t \mid \mathcal{F}_s) = X_s \text{ a.s.}$$

More generally, if u and v are two simple stopping times, we have $X_u = E(X_v \mid \mathcal{F}_u)$ if $u \leqslant v < \infty$ (cf. Section 1.2 for the definition of $\mathcal{F}_u$ and Lemma 1.9).

If X is a real process, one says that X is a supermartingale (resp. a submartingale) if the Doléans function is negative (resp. positive). The definition then gives for a supermartingale

$$E(X_t \mid \mathcal{F}_s) \leqslant X_s \text{ a.s.} \qquad \text{if} \quad s \leqslant t,$$

and for a submartingale,

$$E(X_t \mid \mathcal{F}_s) \geqslant X_s \text{ a.s.} \qquad \text{if} \quad s \leqslant t.$$

Proposition *Every $\mathbf{R}^n$-valued martingale (resp. real-valued submartingale) which is right continuous in probability has a cadlag modification, which is a martingale (resp. a real submartingale) with respect to the σ-algebras $(\mathcal{F}_{t^+})$.*

Proof The proposition is an immediate consequence of the theorem of Section 1.16.

1.18 Examples of Submartingales

Proposition (1) *Let M be a Banach-valued martingale. Then $\|M\|$ is a submartingale.*

(2) *For every convex real increasing function g, defined on the real line,*

and every real submartingale X, $g(X)$ is a submartingale as soon as $g(X_t)$ is integrable for every t.

Proof (1) By writing

$$\|M_s\| \leqslant \|E((M_s - M_t) \mid \mathcal{F}_s)\| + \|E(M_t \mid \mathcal{F}_s)\| = \|E(M_t \mid \mathcal{F}_s)\| \qquad \text{for} \quad s < t,$$

and using the relation

$$\left\| \int_F E(M_t \mid \mathcal{F}_s)\, dP \right\| = \left\| \int_F M_t\, dP \right\| \leqslant \int_F \|M_t\|\, dP$$
$$= \int_F E(\|M\|_t \mid \mathcal{F}_s)\, dP \qquad \forall F \in \mathcal{F}_s,$$

which implies that

$$\|E(M_t \mid \mathcal{F}_s)\| \leqslant E(\|M_t\| \mid \mathcal{F}_s) \text{ a.s.}$$

we get immediately that $\|M\|$ is a submartingale.

(2) The second statement is an immediate consequence of the Jensen inequality: let Y be an element of $L^1(\Omega, \mathcal{F}, P)$ and $\mathcal{G}$ a sub-σ-algebra of $\mathcal{F}$; let f be a convex real positive function defined on the real line; we then have

$$f[E(Y \mid \mathcal{G})] \leqslant E[f(Y) \mid \mathcal{G}].$$

This inequality is obvious if $f(x) = ax + b$; thus the same inequality holds in the general case since a convex function is the supremum of a family $(f_n)_{n>0}$ of functions of the form

$$f_n(x) = a_n x + b_n.$$

1.19 Brownian and Poisson Processes as Martingales and Submartingales

Let X be a brownian motion in $\mathbf{R}^n$ with variance parameter σ. By property [I] (see Section 1.5), the Doléans function x of X is clearly

$$x(F \times]s, t]) = P(F)E(X_t - X_s) = 0;$$

therefore X is a martingale.

Considering the real positive process $Y = \|X\|^2$, which is a submartingale according to Section 1.18, we see that its Doléans function y is, in view of property [B] (see Section 1.5),

$$y(F \times]s, t]) = P(F)\sigma^2(t - s).$$

This function clearly has a (unique) σ-additive extension to $\mathcal{P}$, which is the restriction to $\mathcal{P}$ of $\sigma^2 \cdot P \otimes l$ (defined on $\mathcal{F} \otimes \mathcal{B}_T$), where l is the Lebesque measure on $\mathcal{B}_T$.

Now let N be a Poisson process with parameter λ. Since it is a positive process with increasing paths, it is clearly a submartingale. Moreover, its Doléans function (using property [I] as above for brownian motion) is given by

$$n(F \times]s,t]) = P(F)E(N_t - N_s) = \lambda(t-s)P(F).$$

This Doléans function extends into a measure which is the restriction to $\mathcal{P}$ of the product measure $\lambda P \otimes l$.

It can be noticed immediately that the process $(N_t - \lambda t)$ has a zero Doléans function and is therefore a martingale.

1.20 Doléans Measure of a Positive Submartingale

In this section we shall use the following simple version of a more general theorem that will be studied in Section 8.

Proposition *Let X be a positive submartingale on $T := [0, t_m]$ with the property that for every*

$$t \in [0, t_m[, \quad \lim_{s \downarrow t,\, s > t} E(X_s - X_t) = 0.$$

Then the Doléans function of X has a σ-additive extension.

Proof We begin with a remark. Every set $A \in \mathcal{A}$ with $A \subset]0, t_m]$ is included in the set $]u, t_m]$ defined by

$$u(\omega) := \inf\{t : (\omega, t) \in A\} \wedge t_m.$$

It is easily seen that $]u, t_m] \in \mathcal{A}$, and if $A \subset H \times]0, t_m]$, $u(\omega) = t_m$ for every $\omega \notin H$. Therefore

$$X_{t_m} - X_u = 1_H(X_{t_m} - X_u).$$

Let us call x the Doléans function of X. Since $x(]u, t_m]) = E(X_{t_m} - X_u)$, we derive from the positivity of x

$$x(A) \leqslant E(X_{t_m} - X_u) = E\left[1_H(X_{t_m} - X_u)\right]. \tag{1.20.1}$$

Using this remark, we now prove that the positive additive function x has a σ-additive extension to $\mathcal{P}$. For this it is enough to show that for any decreasing sequence (A_n), $A_n \subset \mathcal{A}$ with $\bigcap_n A_n = \varnothing$ one has $\lim_n x(A_n) = 0$. With every predictable rectangle $F \times]s, t]$, we associate the sets $F \times]s + (1/n), t] \subset F \times [s + (1/n), t] \subset F \times]s, t]$, $n \geqslant 1$, and from the hypothesis in the proposition, we notice that

$$\lim_{n \to \infty} x\left(F \times \left]s + \frac{1}{n}, t\right]\right) = x(F \times]s, t]).$$

Therefore, for every given $\epsilon > 0$ we can find a sequence (A'_n) in $\mathcal{A}$ and a sequence (C_n) of subsets of $\Omega \times]0, t_m]$ with the following properties for every n:

$$A_n \supset C_n \supset A_n', \tag{1.20.2}$$

$$x(A_n - A_n') \leqslant \epsilon 2^{-n}; \tag{1.20.3}$$

$C_n(\omega) := \{t : (\omega, t) \in C_n\}$ is compact for every $\omega \in \Omega$. The hypothesis $\bigcap_n A_n = \emptyset$ and the property of C_n imply that

$$\bigcap_n H_n = \emptyset, \tag{1.20.4}$$

where

$$H_n := \left\{\omega : \bigcap_{k \leqslant n} C_k(\omega) \neq \emptyset\right\}.$$

Since

$$\bigcap_{k \leqslant n} A_k' \subset H_n \times]0, t_m],$$

the above remark shows (inequality (1.20.1)) that

$$x\left(\bigcap_{k \leqslant n} A_k'\right) \leqslant \int_{H_n} X_{t_m}\, dP.$$

From (1.20.4) we therefore derive

$$\lim_{n \to \infty} x\left(\bigcap_{k \leqslant n} A_k'\right) = 0,$$

while

$$x\left(A_n - \bigcap_{k \leqslant n} A_k'\right) \leqslant \epsilon \sum_{k \leqslant n} 2^{-k} \leqslant 2\epsilon.$$

Therefore

$$\lim_n x(A_n) \leqslant 2\epsilon \qquad \text{for all} \quad \epsilon > 0,$$

and this proves the proposition.

OPTIONAL PROCESSES

1.21 Optional Sets and Processes (Definitions)

Let $\mathcal{O}$ be the σ-algebra of subsets of $\Omega' = \Omega \times T'$ generated by the stochastic intervals $]0, u[$ for all the stopping times u. This σ-algebra is called the σ-algebra of the *optional sets*.[1] One says that X is an optional process if X is measurable with respect to this σ-algebra $\mathcal{O}$.

[1] Formerly called *well-measurable sets*.

Of course the σ-algebra $\mathcal{P}$ of the predictable sets is contained in the σ-algebra $\mathcal{O}$ of the optional sets (in view of the following equality: $]0, u] = \bigcap_{n>0}]0, u + (1/n)[$). Conversely, let $\mathcal{O}'$ be a σ-algebra such that $\mathcal{P}$ is contained in $\mathcal{O}'$ and such that for every stopping time u, $[u]$ belongs to $\mathcal{O}'$; then $\mathcal{O}$ is contained in $\mathcal{O}'$.

1.22 Right Continuous and Optional Processes

Proposition *Let* **H** *be a Banach space; let X be an* **H**-*valued adapted cadlag process; then X is an optional process with respect to the family $(\mathcal{F}_{t^+})_{t\in T}$.*

Proof (1) Let us assume that $\mathcal{F}_t = \mathcal{F}_{t^+}$ for all $t \in T$. We prove first that $Y := X_u 1_{[u, t[}$ is an optional process if u is a stopping time $\leqslant t$; X being adapted, X_u is an $\mathcal{F}_u$-measurable (see Exercise 1) random variable, and it is therefore sufficient to consider the case in which X_u is an $\mathcal{F}_u$-simple random variable. We can thus assume that

$$X_u := \sum_{i\in I} a_i 1_{F(i)},$$

with $a_i \in H$ and $F(i) \in \mathcal{F}_u$ for each element i of I; in this case

$$Y := \sum_{i\in I} a_i 1_{F(i)} 1_{[u, t[}.$$

If we put $u(i) := u$ if $\omega \in F(i)$ and $u(i) := t$ if $\omega \notin F(i)$, we have

$$Y = \sum_{i\in I} a_i 1_{[u(i), t[},$$

and this proves that Y is an optimal process, [$u(i)$ being a stopping time for every i in I (see Exercise 5)].

(2) Let us now consider the general case. For every integer $n > 0$ let $(u(n,k))_{k>0}$ be the sequence of stopping times (with respect to the family $(\mathcal{F}_{t^+})_{t\in T}$ (cf. Section 1.6)) defined by $u(n, 0) = 0$ and

$$\begin{aligned} u(n,k+1) &:= \inf\{t : t \geqslant u(n,k), \|X_t - X_{u(n,k)}\| > 1/n\}, \\ u(n,k+1) &:= \infty \qquad \text{if the set is empty.} \end{aligned}$$

Let X^n be the process defined by $X_t^n := X_{u(n,k)}$ for $u(n,k) \leqslant t < u(n,k+1)$. The process X^n is well defined because $[u(n,k) < \infty] \downarrow_{k\to\infty} \emptyset$ and is optional (cf. (1)); but the sequence $(X^n)_{n>0}$ converges uniformly to the process X; thus X is an optional process.

EXTENSIONS AND EXERCISES

1 If u is a stopping time with $u(\omega) \in T$ a.s., then for every process X on T we can define the random variable X_u by $X_u(\omega) = X_{u(\omega)}(\omega)$ a.s. *If X is* *adapted and right continuous, X_u is $\mathcal{F}_u$-measurable when $(\mathcal{F}_t)$ is right continuous.*

2 *An example of nonpredictable stopping time.* We take for Ω the two points set $\{\omega_0, \omega_1\}$. For all $t < 1$, $\mathcal{F}_t$ is the σ-algebra $\{\Omega, \emptyset\}$ and for $t \geqslant 1$, $\mathcal{F}_t$ is the σ-algebra $\mathfrak{P}(\Omega)$. Show that the real function u, defined by $u(\omega_0) = 1$ and $u(\omega_1) = 2$, is a *nonpredictable* stopping time. (We assume that $0 \neq P(\omega_0) \neq 1$.)

3 *If u is a stopping time and ξ is an $\mathcal{F}_u$-measurable random variable, then $\xi \cdot 1_{]u, \infty[}$ is a predictable process.*

4 *If X is a predictable process and u is a stopping time, then $(X_{u \wedge t})_{t \in T}$ is predictable.*

5 If u is a stopping time and $F \in \mathcal{F}_u$, the random variable

$$u_F(\omega) := \begin{cases} u(\omega) & \text{if } \omega \in F \\ +\infty & \text{if } \omega \notin F \end{cases}$$

is a stopping time.

6 Let X be a cadlag adapted process on T. For every $\delta > 0$, the random variable u defined by (we assume $(\mathcal{F}_t = (\mathcal{F}_{t^+}))$

$$u(\omega) := \inf\{ t : \|X_t(\omega) - X_{t^-}(\omega)\| > \delta \}$$

is a stopping time. (Consider an increasing family (S_n) of finite totally ordered subsets of $T: S_n := \{t_1^n < t_2^n < \cdots < t_{k(n)}^n\}$, the union of which is dense in T, and set

$$u_n := \inf_j \{ t_j^n : \|X_{t_j^n} - X_{t_{j-1}^n}\| > \delta \};$$

the u_n's are stopping times which converge to u.)

7 Let $(\Omega, \mathcal{F}, P, (\mathcal{F}_t)_{t \in T})$ be the probabilized stochastic basis defined by $\Omega := \{a, b\}$ (two-point set), $\mathcal{F} := \mathfrak{P}(\Omega)$ (the family of all the subsets of Ω), $P(\{a\}) := P(\{b\}) := \frac{1}{2}$, $T := [0, 1]$ (unit interval of the real line), $\mathcal{F}_t := (\emptyset, \Omega)$ (trivial σ-algebra) if $t \leqslant \frac{1}{2}$ and $\mathcal{F}_t := \mathcal{F}$ if $t > \frac{1}{2}$.

(a) We define

$$u(a) := 1; \qquad u(b) := \tfrac{1}{2};$$
$$v(a) := 0; \qquad v(b) := \tfrac{1}{2}.$$

Are u and v stopping times?

(b) Is it true that $\mathcal{F}_t = \mathcal{F}_t^+$ for each element t of T?

(c) Are u and v stopping times for the family $(\mathcal{F}_{t^+})_{t \in T}$?

(d) Is $[0, u[$ a predictable set?

(e) Is $X := 1_{[0, u[}$ a predictable process? An adapted process?

(f) Let $(w_n)_{n>0}$ be the sequence of random variables defined by $w_n(a) = 1$ and $w_n(b) = \frac{1}{2} + 1/n$ for each integer n. We put $w = \inf_{n>0} w_n$. Is w_n a stopping time? Is w a stopping time for the family $(\mathcal{F}_t)_{t \in T}$? For the family $(\mathcal{F}_{t^+})_{t \in T}$?

8 We define $(\Omega, \mathcal{F}, P)$ as in Exercise 7. Let Y be the process defined by $Y_t(b) := 0$ for every element t of T, $Y_t(a) := 0$ for $t \leqslant \frac{1}{2}$, and $Y_t(a) := t - \frac{1}{2}$ for $t > \frac{1}{2}$. For every element t of T let $\mathcal{G}_t$ be the σ-algebra generated by the random variables $(Y_s)_{s \leqslant t}$ (i.e., the smallest σ-algebra for which these random variables are measurable). Compare these σ-algebras $(\mathcal{G}_t)_{t \in T}$ and the σ-algebras $(\mathcal{F}_t)_{t \in T}$ of Exercise 7. Is the process Y adapted or predictable with respect to the stochastic basis $(\Omega, \mathcal{F}, P, (\mathcal{F}_t)_{t \in T})$? In this case $(\Omega, \mathcal{F}, P, (\mathcal{G}_t)_{t \in T})$ is often called the canonical stochastic basis of the process Y.

9 Let $(u(n))_{n>0}$ be an increasing sequence of stopping times. We put $u := \sup_{n>0} u(n)$. Is u a stopping time?

10 Let $(\Omega, \mathcal{F}, P, (\mathcal{F}_t)_{t \in T})$ be a stochastic basis with $T = [0, 1]$. Are the following assertions true or false (to show that one of the following assertions is false, one may use Exercise 7):

(a) Let u be a T-valued random variable defined on $(\Omega, \mathcal{F}, P)$; then u is a stopping time if and only if for each element t of T the set $A_t := \{\omega : u(\omega) < t\}$ belongs to $\mathcal{F}_t$.

(b) Let u be a T-valued random variable defined on $(\Omega, \mathcal{F}, P)$; then u is a stopping time if and only if for each element t of T the random variable $(u \wedge t)$ is $\mathcal{F}_t$-measurable.

(c) Let u and v be two stopping times with $u \geqslant v$. Let w be the random variable defined by

$$\begin{aligned} w(\omega) &:= 1 && \text{if} \quad u(\omega) > v(\omega), \\ w(\omega) &:= u(\omega) && \text{if} \quad u(\omega) = v(\omega); \end{aligned}$$

then w is a stopping time.

Answer: All the assertions are true when $(\mathcal{F}_t)$ is right continuous and are false otherwise.

2 STOCHASTIC INTEGRAL

In this section we give an elementary definition of the stochastic integral for a wide class of Banach-valued predictable processes with respect to another wide class of Banach-valued processes.

The extent to which this last class is general will be evaluated with exactitude in Section 10, where this class will be found to contain all the so-called Hilbert-valued "semimartingales." We shall start here with an extremely simple definition of the considered processes so that we do not need any preliminary study of martingales, semimartingales, etc. to perform the construction of the stochastic integral.

It should be noted that our considering Banach-valued processes does not introduce any difficulty into the theory. The reader who is reluctant to hear of Banach-valued processes has only to read "real valued" everywhere instead of "Banach valued," and to replace **L**, **J**, and **K** by **R**!

2.1 The Problem of Stochastic Integration

Throughout this section $T = [0, t_m]$, $t_m < \infty$, and we consider a complete right-continuous probabilized stochastic basis $(\Omega, \mathcal{F}, P, (\mathcal{F}_t)_{t \in T})$ (cf. Section 1.1), three Banach spaces **L**, **J**, and **K**, where **L** is a Banach space imbedded in the space $\mathcal{L}(\mathbf{J}; \mathbf{K})$ of bounded linear operators from **J** into **K**. The norms in **L**, **J**, and **K** will be denoted by $\|\cdot\|_{\mathbf{L}}$, $\|\cdot\|_{\mathbf{J}}$, and $\|\cdot\|_{\mathbf{K}}$, respectively. We assume that $\|u\|_{\mathbf{L}} \geqslant \sup_{\|x\|_J \leqslant 1} \|u(x)\|_{\mathbf{K}}$.

What is the problem of stochastic integration?

Let Y be an **L**-valued process (usually Y is a predictable process) and X be a **J**-valued process defined up to modification; then, the problem is to

(1) define for each element t of T, the random variable

$$Z_t = \int_0^t Y_s\, dX_s = \int 1_{]0,t]}(s)\, Y_s\, dX_s,$$

and

(2) study the process $(Z_t)_{t \in T}$ thus defined up to modification.

The processes X considered here have a cadlag modification; we shall also write X for this cadlag modification defined up to P-equivalence. It is natural then to define $Z_t(\omega)$ as the usual integral of the **L**-valued sample function $s \rightsquigarrow Y_s(\omega)$ with respect to the "measure" $dX_s(\omega)$ (ω being fixed), if such an integral exists.

Actually, this construction is not possible for the general case. The reason for this is that many processes, for example, real brownian motion, have paths $t \rightsquigarrow X_t(\omega)$ with unbounded variation, in which case $dX_t(\omega)$ (ω being fixed) does not define a measure.

The construction that we now describe is not the most general but is very elementary and will lead, through several steps, to the definition of the stochastic integral with respect to a very wide class of processes.

2.2 $\mathcal{Q}$-Simple Processes; Notation $\mathcal{E}(\mathbf{L})$

$\mathcal{E}(\mathbf{L})$ will denote the vector space of the **L**-valued and $\mathcal{Q}$-simple processes, i.e., the processes Y such that $Y := \sum_{i \in I} a_i 1_{A(i)}$ where $(a_i)_{i \in I}$ is a finite family of elements of **L** and $(A(i))_{i \in I}$ is a finite family of elements of $\mathcal{Q}$.

According to Section 1.8, we may assume that in the above expression of Y the sets $(A(i))_{i \in I}$ are disjoint and belong to $\mathcal{R}$.

In this case, even though $dX_t(\omega)$ does not define a *Stieltjes* measure on T, there is an immediate natural meaning for

$$Z_t(\omega) := \int 1_{]0,t]}(s)\, Y_s(\omega)\, dX_s(\omega), \tag{2.2.1}$$

which reduces to its usual one when $dX_s(\omega)$ is a ($\mathbf{J}$-valued) measure on T. This is

$$\int Y_s(\omega)\, dX_s(\omega) = 1_F(\omega) a(X_t(\omega) - X_s(\omega)) \tag{2.2.2}$$

for $Y := a1_{F\times]s,t]}$, $a \in \mathbf{L}$.

For a general, simple process Y, the definition follows by linearity.

It is clear from formula (2.2.2) that the process Z is cadlag as soon as X has this property. If X is defined up to a modification, the same is true for Z, defined by (2.2.1).

Then the stochastic integral $\int_{]0,t]} Y\,dX$ is the linear mapping defined on $\mathcal{E}(\mathbf{L})$ with values in $L^0_{\mathbf{K}}(\Omega, \mathcal{F}, P)$ such that, for each element $A = F \times]s,u]$ of $\mathcal{R}$ and each element a of $\mathbf{L}$, if $Y = a1_A$, we have

$$\int_{]0,t]} Y\,dX = \int_{]0,t]} a1_A\,dX = 1_F a(X_{t\wedge u} - X_{s\wedge u}). \tag{2.2.3}$$

The problem is to extend the mapping $Y \mapsto \int_{]0,t]} Y\,dX$ to a class of processes larger than the class of $\mathcal{Q}$-simple processes.

2.3 A First Extension

Definition Let X be a $\mathbf{J}$-valued process defined up to a modification. It will be called an $(\mathbf{L},\mathbf{J},\mathbf{K})$-$\mathbf{L}^2$-primitive process (or briefly, a L^2-primitive process if the triple $(\mathbf{L},\mathbf{J},\mathbf{K})$ is clear from the context), if the following property holds:

[i] there exists a finite positive measure α on predictable sets, vanishing on evanescent sets[2] and such that, for every $\mathbf{L}$-valued $\mathcal{Q}$-simple process Y, we have

$$E\left(\left\|\int_T Y\,dX\right\|_K^2\right) \leqslant \int_{\Omega'} \|Y\|_{\mathbf{L}}^2\, d\alpha < \infty.$$

We say in this case that the measure α dominates X (for the bilinear mapping $\mathbf{L}\times\mathbf{J}\to\mathbf{K}$).

It should be emphasized that the domination property for X depends on the considered Banach spaces $\mathbf{J}$ and $\mathbf{K}$ (see Exercise 1)).

[2] This condition on α could be released, but this would introduce no more generality. Later, we will call *admissible* a measure α with this property.

In this case, the mapping $Y \rightsquigarrow \int Y\,dX$ defined on $\mathcal{E}(\mathbf{L})$ with values in $L^2_{\mathbf{K}}(\Omega, \mathcal{F}, P)$ is uniformly continuous if we consider $\mathcal{E}(\mathbf{L})$ as a subspace of $L^2_{\mathbf{L}}(\Omega', \mathcal{P}, \alpha)$; then there is a unique extension of this mapping into a linear continuous mapping from $L^2_{\mathbf{L}}(\Omega', \mathcal{P}, \alpha)$ into $L^2_{\mathbf{K}}(\Omega, \mathcal{F}, P)$ (the space $\mathcal{E}(\mathbf{L})$ being dense in $L^2_{\mathbf{L}}(\Omega', \mathcal{P}, \alpha)$). The image of a process Y belonging to $L^2_{\mathbf{L}}(\Omega', \mathcal{P}, \alpha)$ by this mapping will be denoted by $\int Y\,dX$ and will be called the stochastic integral of the process Y with respect to the process X.

Remarks One may feel unhappy that this definition of $\int Y\,dX$ depends on a measure α, which is by no means unique in general. It should therefore be noticed that the extension procedure, which we shall further develop, will provide us with an extension to a class of processes which does not depend upon a particular dominating measure. Moreover, the following observations can be made now:

(1) On the set $\mathcal{E}(\mathbf{L})$ the mapping $Y \rightsquigarrow \Phi(Y) := [E\|\int Y\,dX\|^2_{\mathbf{K}}]^{1/2}$ defines clearly a hilbertian seminorm Φ. For every dominating measure α, the corresponding seminorm $\|\ \|_\alpha$ in $L^2_{\mathbf{L}}(\Omega', \mathcal{P}, \alpha)$ is such that $\Phi(Y) \leqslant \|Y\|_\alpha$ for all $Y \in \mathcal{E}(\mathbf{L})$, and the same inequality holds by continuity for all $Y \in L^2_{\mathbf{L}}(\Omega', \mathcal{P}, \alpha)$.

As a consequence, *for two dominating measures α_1 and α_2, the above extensions of $Y \rightsquigarrow \int Y\,dX$ to $L^2_{\mathbf{L}}(\Omega', \mathcal{P}, \alpha_1)$ and $L^2_{\mathbf{L}}(\Omega', \mathcal{P}, \alpha_2)$ coincide on $L^2_{\mathbf{L}}(\Omega', \mathcal{P}, \alpha_1) \cap L^2_{\mathbf{L}}(\Omega', \mathcal{P}, \alpha_2)$.* If we call $L^2(X; \mathbf{L}; \mathbf{K})$ the set of predictable processes defined by $L^2(X; \mathbf{L}; \mathbf{K}) = \bigcup_{\alpha \in D} L^2_{\mathbf{L}}(\Omega', \mathcal{P}, \alpha)$, where D is the set of dominating measures for X (related to the spaces $\mathbf{L}$ and $\mathbf{K}$), we have then defined an extension of $Y \rightsquigarrow \int Y\,dX$ from $\mathcal{E}(\mathbf{L})$ to $L^2(X; \mathbf{L}; \mathbf{K})$ which is a continuous mapping from $L^2(X; \mathbf{L}; \mathbf{K})$ endowed with the seminorm Φ into $L^2_{\mathbf{K}}(\Omega, \mathcal{F}, \mathcal{P})$.

(2) It will be seen in Section 2.6 that when X is a martingale such that $E\|X_t\|^2 < \infty$ for all $t \in T$, and $\mathbf{K}$ and $\mathbf{J}$ are Hilbert spaces, there exists, among the dominating measures, one—say α_0—which is the smallest (for the ordering of positive measures). In this case $L^2(X; \mathbf{L}; \mathbf{K}) = L^2_{\mathbf{L}}(\Omega', \mathcal{P}, \alpha_0)$. Moreover, in this special case, if $\mathbf{L}$, $\mathbf{J}$ and $\mathbf{K}$ are the real line, the seminorms Φ and $\|\ \|_{\alpha_0}$ coincide.

2.4 The Stochastic Integral Process

Let X be a process defined up to a modification which satisfies the condition [i] (Section 2.3) and Y be a process which belongs to $L^2(X; \mathbf{L}, \mathbf{K})$. For every t in T, we can define the random variable Z_t by

$$Z_t := \int 1_{]0,t]} Y\,dX.$$

This defines the process Z up to modification. We shall now study this process and show, in particular, that if X is cadlag, it has a unique (up to P-equivalence) cadlag modification. This cadlag process will be called the *stochastic integral process of Y with respect to X* and will be denoted by

$$\left(\int Y\, dX\right) \quad \text{or} \quad \left(\int_{]0,\,t]} Y\, dX\right)_{t\in T}$$

Before proving the existence of a cadlag modification (in Section 2.5), we first state a property of Z which does not depend on this property:

Proposition *Let X be a real L^2-primitive process (where* $\mathbf{L} = \mathbf{J} = \mathbf{K} = \mathbf{R}$), *with dominating measure α. Then for every $Y \in L^2(\Omega', \mathcal{P}, \alpha)$, the integral process $Z := (\int Y\, dX)$ admits the measure $A \rightsquigarrow \int_A |Y|^2\, d\alpha$ as one dominating measure (for the real multiplication).*

In the case of vector-valued processes, the analogous statement holds but is more complicated in its formulation. For the ease of the reader, we do not consider it until Exercise 2 at the end of this section.

Proof For every pair (Y, U) of real simple processes, it can easily be shown that $\int U\, dZ = \int UY\, dX$. Indeed, this statement is trivial for $U = 1_{F\times]0,\,u]}$ and $Y = 1_{G\times]0,\,v]}$ and follows by linearity for any couple of real simple processes. Therefore, by continuity, it holds for every $Y \in L^2(\Omega', \mathcal{P}, \alpha)$. Hence, we derive

$$E\left|\int U\, dZ\right|^2 \leqslant \int |U|^2 |Y|^2\, d\alpha.$$

2.5 Uniform Approximation and Dominated Convergence Theorems

We consider the hypothesis and notations given in Sections 2.1, 2.2, and 2.3. Moreover, we assume that the family $(\mathcal{F}_t)_{t\in T}$ is right continuous, the basis $(\Omega, \mathcal{F}, P, (\mathcal{F}_t)_{t\in T})$ is complete, and X is a cadlag adapted process. Let $(Y^n)_{n>0}$ be a sequence of $\mathcal{Q}$-simple processes such that for any integer n

$$\int \|Y - Y^n\|_{\mathbf{L}}^2\, d\alpha \leqslant 8^{-n}.$$

For each integer n let Z^n be the cadlag process defined by the formula (2.2.3) applied to the process Y^n. For each integer n, we set

$$u(n) = \inf\{t : \|Z_t^n - Z_t^{n+1}\|_{\mathbf{K}} > 2^{-n}\} \wedge t_m.$$

Let $G(n)$ be the set defined by

$$G(n) = \{\omega : (u(n))(\omega) < t_m\}.$$

For every simple stopping time v (i.e., with finitely many values) the processes $1_{]0,v]}Y^n$ belong to $\mathcal{E}(\mathbf{L})$ and from the definition we have

$$Z_v^n = \int 1_{]0,v]} Y^n \, dX \text{ a.s.}$$

Therefore,

$$E(\|Z_v^n - Z_v^{n+1}\|_{\mathbf{K}}^2) = E\left(\left\|\int 1_{]0,v]}(Y^n - Y^{n+1})\, dX\right\|^2\right)$$

$$\leqslant \int 1_{]0,v]} \|Y^n - Y^{n+1}\|_{\mathbf{L}}^2 \, d\alpha \leqslant 2 \cdot 8^{-n}.$$

For a general stopping time, the same inequality follows from the possibility of considering such a stopping time as the decreasing limit of a sequence of simple stopping times (cf. Section 1.11(1)) and from the right continuity of Z^n. Thus, this inequality is satisfied for $v = u(n)$, and we have

$$2 \cdot 8^{-n} \geqslant E(\|Z_{u(n)}^n - Z_{u(n)}^{n+1}\|_{\mathbf{K}}^2) \geqslant 4^{-n} P[G(n)].$$

Then $P[G(n)] \leqslant 2 \cdot 2^{-n}$ and $P(G) = 0$ where $G := \bigcap_{k>0}\{\bigcup_{n \geqslant k} G(n)\}$. Therefore for every $\omega \notin G$, there exists an integer k such that for any integer $n \geqslant k$, $\sup_{t<t_m} \|Z_t^n - Z_t^{n+1}\| \leqslant 2^{-n}$. This means that for every element ω of $\Omega \backslash G$, the sequence $(Z_t^n(\omega))_{n>0}$ is a Cauchy sequence which converges uniformly on $T - \{t_m\}$ to a function $\hat{Z}_t(\omega)$. As $Z_{t_m}^n$ converges in $L_{\mathbf{K}}^2(\Omega, \mathcal{F}, P)$ to Z_{t_m}, there actually exists a P-null set G' such that for $\omega \in \Omega - G'$, this convergence holds uniformly on T and the process $\hat{Z}$ is a modification of the process Z. Thus, we have proved that *if X has a modification which is a cadlag adapted process, the same property holds for the process Z.* The definition at the beginning of Section 2.4 is therefore justified and the above argument shows

Theorem *Let X be a cadlag process which satisfies the property* [i] (*Section* 2.3). *Let $(Y_n)_{n>0}$ be a sequence of* **L**-*valued processes which converges to Y in the following sense: for each integer n,*

$$\int \|Y - Y_n\|_{\mathbf{L}}^2 \, d\alpha \leqslant \epsilon_n \delta_n^2,$$

$$\textit{with } \sum_n \epsilon_n < \infty, \textit{ and } \lim_n \downarrow \delta_n = 0.$$

For every integer n, let Z^n be a cadlag process, which is modification of the stochastic integral process $\int Y_n \, dX$; then for P-almost all ω, the paths $Z_n(\omega, \cdot)$ of the sequence $(Z_n)_{n>0}$ converge uniformly on T to the paths of a cadlag process, which is a modification of the stochastic integral process $Z := (\int Y \, dX)$.

Corollary *Let $(Y_n)_{n>0}$ be a sequence which converges α-almost everywhere to Y for some dominating measure α. Then there exists a subsequence*

$(Y_{n(k)})_{k>0}$ such that the corresponding subsequence $(Z_{n(k)})_{k>0}$ of the cadlag integral processes converge to the cadlag integral process $Z = (\int Y\,dX)$ uniformly for P-almost all paths ω.

Applications This theorem is very useful in proving many properties. We give some examples (details of the proofs are left to the reader):

A.1 If X has a modification which is a continuous (or predictable, etc.) process, it is the same for the integral process $(\int Y\,dX)$.

A.2 If u is a stopping time, and if α is a dominating measure for X, $1_{[0,u]}\alpha$ is clearly a dominating measure for $\overline{X}^u$, which is thus a primitive process. The stochastic integral process $(\int Y\,dX)$ stopped at u is the same as the integral process of Y with respect to the process X stopped at u.

A.3 If u is a stopping time and Y is predictable, we have

$$Z_u - Z_{u^-} = Y_u(X_u - X_{u^-}),$$

where Z is the cadlag process $(\int Y\,dX)$ and X is a cadlag adapted process.

All these properties are obvious if Y is an $\mathcal{Q}$-simple process; they are true in the general case from the dominated convergence theorem above. (For A.3, we use a starting approximation for Y as in Section 1.11.)

2.6 Special Cases: Integrating with Respect to Brownian Motion and More Generally Square Integrable Martingales

Returning to the considerations in Section 1.19, it can easily be seen that for a real brownian motion X with variance parameter σ and $\mathbf{L} = \mathbf{R}$, the measure $\sigma^2 P \otimes l$ can be taken as the measure α in the formula (2.3)[i].

Assuming, moreover, that Y and X are real, we can further write:

$$E\left(\int |Y\,dX|^2\right) = \int |Y|^2\,d\alpha. \tag{2.6.1}$$

This proves that the mapping $Y \mapsto \int Y\,dX$ is an isometry from $L^2(\Omega', \mathcal{P}, \alpha)$ into $L^2(\Omega, \mathcal{F}, P)$ and at the same time shows the existence of a smaller dominating measure for the process X, namely the measure $\sigma^2 P \otimes l$.

We now prove a similar property for Hilbert-valued processes with independent increments with zero mean and finite second order moments, and more generally for all Hilbert-valued martingales M on $[0, t_m]$ with the property $E\|M_{t_m}\|^2 < \infty$, which we call *square integrable*, when $\mathbf{K}$ is moreover a Hilbert space.

Indeed: writing

$$Y = \sum_i a_i 1_{F_i \times]s_i t_i]}$$

where the rectangles $F_i \times]s_i t_i]$ are assumed to be disjoint, we easily get

$$E\left\{1_{F_i \cap F_j}\langle a_i(M_{t_i} - M_{s_i}), a_j(M_{t_j} - M_{s_j})\rangle_{\mathbf{K}}\right\} = 0$$

from the martingale property when $i \neq j$, and by denoting by $\langle\ ,\ \rangle_{\mathbf{K}}$ the scalar product in **K**. This equality follows from the fact that either $F_i \cap F_j = 0$ or $1_{F_i \cap F_j} a_i(M_{t_i} - M_{s_i})$ is $\mathcal{F}_{s_j}$-measurable as soon as $s_j \geqslant t_i$. Therefore

$$E\left\|\int Y\,dM\right\|^2 = \sum_i E\left\{1_{F_i}\|a_i(M_{t_i} - M_{s_i})\|_{\mathbf{K}}^2\right\}. \tag{2.6.2}$$

If M in particular is with independent centered increments, we have

$$E\left(1_{F_i}\|M_{t_i} - M_{s_i}\|^2\right) = P(F_i)(f(t_i) - f(s_i)),$$

where f is an increasing function. Calling μ the Stieltjes measure of f, (2.6.2) immediately gives

$$E\left\|\int Y\,dM\right\|^2 \leqslant \int \|Y\|^2\,d(P \otimes \mu) \tag{2.6.3}$$

with equality if the processes Y and M are real. In this case $P \otimes \mu$ is therefore the smallest dominating measure.

Coming back to a general Hilbert-valued cadlag square integrable martingale, we remark that for every predictable rectangle $F \times]s, t]$,

$$E\left(1_F\|M_t - M_s\|^2\right) = E\left[1_F\left(\|M_t\|^2 - \|M_s\|^2\right)\right] - 2E\,1_F\langle M_s, M_t - M_\wedge\rangle_J,$$

and from the martingale property,

$$E\left(1_F\|M_t - M_s\|^2\right) = E\left[1_F\left(\|M_t\|^2 - \|M_s\|^2\right)\right]. \tag{2.6.4}$$

We now recall that $\|M\|^2$ is a positive submartingale (see Section 1.18). In particular, $E\|M_t\|^2 \leqslant E\|M_{t_m}\|^2$ for all $t \leqslant t_m$. Let us assume that the mapping $t \rightsquigarrow M_t$ is right continuous from T into $L^2_{\mathbf{J}}(\Omega, \mathcal{F}, P)$. The proposition given in Section 1.20 shows that the Doléans measure of $\|M\|^2$ has a positive σ-additive extension to $\mathcal{P}$, which we denote by α.

For every $\mathcal{Q}$-simple process Y equalities (2.6.2) and (2.6.4) therefore give

$$E\left(\left\|\int_T Y\,dM\right\|^2\right) \leqslant \int \|Y\|^2\,d\alpha,$$

with equality if Y and M are real.

We have thus proved the first statement of the following proposition.

Proposition *Every* **J**-*valued square integrable martingale M such that $t \rightsquigarrow M_t$ is a right continuous*[6] *mapping from T into $L^2_{\mathbf{J}}(\Omega, \mathcal{F}, P)$, is an* (**L**, **J**, **K**)-

[6] In Section 8, it will be seen that if M is a square integrable martingale, $(\|M_t\|^2)_{t \leqslant t_m}$ is equi-integrable. Therefore, as a consequence of (7.6.4), if M is a square integrable cadlag martingale, the mapping $t \rightsquigarrow M_t$ is $L^2_{\mathbf{J}}(\Omega, \mathcal{F}, P)$-continuous.

L^2-primitive process for all Hilbert spaces **J** *and* **K** *for* $\mathbf{L} = \mathfrak{L}(\mathbf{J}, \mathbf{K})$. *For every predictable* **L**-*valued process* Y, *integrable in the sense of Section* 2.3, *the process* $(\int Y\,dM)$ *is a square integrable martingale. In particular, the stopped process* $(M_{u\wedge t})_{t\in[0,t_m]} = M_0 + (\int 1_{]0,u]}\,dM)$ *is a square integrable martingale for every stopping time* u.

Proof The first statement has been proved above.

Let us recall that a martingale M is characterized (see Section 1.17) by the property

$$E\{1_F(M_t - M_s)\} = 0$$

for every predictable rectangle $F \times]s, t]$.

For every predictable rectangle G and every $\mathcal{Q}$-simple process Y, we therefore have

$$E\left\{\int_T 1_G Y\,dM\right\} = 0.$$

This can be written as

$$E\left\{\int_T 1_G\,dZ\right\} = 0$$

if Z denotes the integral process $(\int Y\,dM)$. Therefore Z is a martingale which is square integrable because of the inequality (2.3)[i]. Using the continuity of the mapping $Y \rightsquigarrow \int 1_G Y\,dM$, the same property holds for every integrable Y.

2.7 Stochastic Integral with Respect to Poisson Process. Stochastic Integrals and Random Stieltjes Integrals

If we consider the Poisson process N or, more generally, a process X with finite variation, it should be remarked that for every bounded predictable process Y, the Stieltjes integration $\int_0^t Y(s,\omega)\,dN(s,\omega)$, with respect to the variable s, can be performed on every path. An unavoidable question is then, What is the relation between this random Stieltjes integral and the above defined stochastic integral? The answer lies in the following proposition.

Proposition *Let* X *be a* **J**-*valued adapted cadlag process on* $T := [0, t_m]$ *with paths of finite variation on* $[0, t_m]$. *Denoting the variation of* X *on* $[0, t]$ *for all the paths* ω *by* $|X|_t(\omega)$, *we assume that* $E(|X|_{t_m}^2) < \infty$. *Then,*

(1) *X is an L^2-primitive process for every Banach* **K** *and* $\mathbf{L} \subset \mathfrak{L}(J; \mathbf{K})$ *with dominating measure* α *defined by*

$$\int_{\Omega\times T} 1_G\,d\alpha = E\left(|X|_{t_m}\int_{[0,t_m]} 1_G\,d|X|_s\right) \tag{2.7.1}$$

for every predictable set $G \subset \Omega'$, where the integral on the right side of this equality should be understood to be a Stieltjes integral on each path.

(2) *Every process Y in $L^2_L(\Omega', \mathcal{P}, \alpha)$ has a Stieltjes integral with respect to dX on almost all paths, and the process thus defined up to P-equivalence is the same as the stochastic integral process.*

Proof (1) It is immediately seen that (2.7.1) defines a measure α on $(\Omega', \mathcal{P})$, and by applying the Cauchy–Schwarz inequality to the Stieltjes integral $\int_{[0, t_m]} Y_s(\omega)\, dX_s(\omega)$, we obtain

$$\left\| \int_{[0, t_m]} Y_s(\omega)\, dX_s(\omega) \right\|^2 \leqslant |X|_{t_m} \int \| Y_s(\omega) \|^2 \, d|X|_s,$$

and from the definition of α, we can deduce

$$E \left\| \int_{[0, t_m]} Y_s(\cdot)\, dX_s(\cdot) \right\|^2 \leqslant \int_{\Omega'} \| Y \|^2 \, d\alpha, \qquad \forall Y \in \mathcal{E}(\mathbf{L}).$$

(2) For every $Y \in L^2(\Omega', \mathcal{P}, \alpha)$, the inequality

$$E\left(|X|_{t_m} \int_{[0, t_m]} \| Y_s(\omega) \|^2 \, d|X|_s(\omega) \right) < \infty$$

implies the integrability of $\| Y(\omega) \|^2$ with respect to the finite measure $dX(\omega)$ for P-almost all ω. Using the fact then that the stochastic integral with respect to X and the random Stieltjes integral with respect to dX_s coincide on $\mathcal{E}(\mathbf{L})$ and a classical continuity argument, we obtain the second part of the proposition.

2.8 The Stopped Processes X^u and $\bar{X}^u$ when X Is an L^2-Primitive Process

Let X be a cadlag adapted $(\mathbf{L}, \mathbf{J}, \mathbf{K})$-$L^2$-primitive process and u a stopping time. The stopped processes $\bar{X}^u$ and X^u have been defined in Sections 1.12 and 1.13, respectively.

We have already noticed in Section 2.5 (A.2) that $\bar{X}^u$ is an $(\mathbf{L}, \mathbf{J}, \mathbf{K})$-$L^2$-primitive process with dominating measure $1_{]0, u]}\alpha$ as soon as α dominates X.

When the stopping time is predictable, the same reasoning can be used for X^u and rests on noticing that $(\int Y\, dX^u) = (\int 1_{]0, u[} Y\, dX)$ for every $\mathcal{Q}$-simple process Y (notice that $1_{]0, u[} Y$ is predictable when u is). The measure $1_{]0, u]}\alpha$ is then a dominating measure for X^u.

But when u is not predictable, it is not clear that X^u is an L^2-primitive process. We have, however, the following proposition.

Proposition *Let X be a cadlag adapted* $(\mathbf{L}, \mathbf{J}, \mathbf{K})$-*$L^2$-primitive process. Let u be a stopping time such that $E(\|X_u - X_{u^-}\|^2) < \infty$. Then X^u is an* $(\mathbf{L}, \mathbf{J}, \mathbf{K})$-*$L^2$-primitive process.*

Proof From the definition, we immediately see that for every $\mathcal{Q}$-simple process Y,

$$\int_T Y\, dX^u = \int_{]0,u]} Y\, dX - Y_u(X_u - X_{u^-}).$$

Therefore, if α is a dominating measure for X,

$$E\left(\left\|\int_T Y\, dX^u\right\|^2\right) \leqslant 2\int 1_{]0,u]}\|Y\|^2\, d\alpha + 2E\left(\|Y_u\|^2\|X_u - X_{u^-}\|^2\right). \quad (2.8.1)$$

Let us write $\Delta X_u := X_u - X_{u^-}$. If we call α_u the measure on $\mathcal{P}$ defined by

$$\alpha_u(G) := \int 1_G(\omega, u(\omega))\|\Delta X_u(\omega)\|^2 P(d\omega),$$

this measure is finite and we see from the inequality (2.8.1) that

$$E\left(\left\|\int_T Y\, dX^u\right\|^2\right) \leqslant 2\int 1_{]0,u]}\|Y\|^2\, d\alpha + 2\int \|Y\|^2\, d\alpha_u.$$

The measure $2(1_{]0,u]}\alpha + \alpha_u)$ is therefore a dominating measure for X^u.

Corollary *Let* $\mathbf{J}$ *be a Hilbert space. For every cadlag* $\mathbf{J}$-*valued square integrable martingale M, every Hilbert space* $\mathbf{K}$ *and every stopping time u with values in $[0, t_m]$, the process M^u is an* $(\mathbf{L}, \mathbf{J}, \mathbf{K})$-*$L^2$-primitive process.*

Proof The only thing to prove is that $E(\|M_u - M_{u^-}\|^2) < \infty$ for every stopping time u. From the Fatou lemma we may write

$$E(\|M_u - M_{u^-}\|^2) \leqslant \liminf_{n\to\infty} E\Big(\sum_k \|M_{(k+1)2^{-n}\wedge t_m} - M_{k2^{-n}\wedge t_m}\|^2$$
$$\times 1_{\{k2^{-n}<u\leqslant(k+1)2^{-n}\}}\Big)$$
$$\leqslant \liminf_{n\to\infty} E\Big(\sum_k \|M_{(k+1)2^{-n}\wedge t_m} - M_{k2^{-n}\wedge t_m}\|^2\Big).$$

From the martingale property, we derive, as we have already done several times,

$$E\left(\|M_{(k+1)2^{-n}\wedge t_m} - M_{k2^{-n}\wedge t_m}\|^2\right) = E\left(\|M_{(k+1)2^{-n}\wedge t_m}\|^2\right) - E\|M_{k2^{-n}\wedge t_m}\|^2$$

and therefore,

$$E(\|M_u - M_{u^-}\|^2) \leqslant E\left(\|M_{t_m}\|^2\right) - E\left(\|M_0\|^2\right) < \infty.$$

A SECOND EXTENSION—PRIMITIVE PROCESSES

2.9 Primitive Processes

At this point, let us remark that a process X the paths of which have finite variation $|X|_t(\omega)$ on every interval $[0, t]$, defines a random Stieltjes integral. This integral coincides with the already defined stochastic integral when $E(|X|_{t_m}^2) < \infty$. When $E(|X|_{t_m}) = +\infty$, we may notice that if we set

$$u_n := \inf\{t : |X|_t > n\},$$

the stopped processes X^{u_n} are such that

$$E(|X^{u_n}|_{t_m}) < n$$

(this is not necessarily true for the processes $\bar{X}^{u_n}$).

In some sense, this allows us to define the stochastic integral with respect to X "strictly before u_n."

This roughly describes the type of construction we are now heading toward. We shall consider a class of processes which can be "prelocalized into L^2-primitive processes" in the sense of the following definition. This class will contain, in particular, all square integrable martingales and all processes with finite variation, and will have many stability properties. It will turn out in some sense (see Section 12) to be the "widest" one, with respect to which a stochastic integral with "proper continuity properties" can be defined.

Definitions (1) Let X be a $\mathbf{J}$-valued cadlag adapted process on $[0, t_m]$. Denoting by X^{u-} the process X stopped strictly before the stopping time u (see Section 1.13), we call X an $(\mathbf{L}, \mathbf{J}, \mathbf{K})$-primitive process, iff there exists an increasing sequence (u_n) of stopping times with $\lim_n P\{u_n < t_m\} = 0$ such that for every stopping time u with values in $[0, t_m]$, the process $X^{u \wedge u_n-}$ is an $(\mathbf{L}, \mathbf{J}, \mathbf{K})$-$L^2$-primitive (in other words, there exists a finite measure $\alpha_{u \wedge u_n}$ on $\mathcal{P}$ such that for every $\mathbf{L}$-valued $\mathcal{Q}$-simple process Y, we have

$$E\left\{\left\|\int Y\, dX^{u \wedge u_n-}\right\|_{\mathbf{K}}^2\right\} \leqslant \int_{]0,\, u \wedge u_n]} \|Y\|^2\, d\alpha_{u \wedge u_n}.$$

The sequence (u_n) will be called a localizing sequence for X (with respect to $(\mathbf{L}, \mathbf{J}, \mathbf{K})$).

(2) Let $\mathbf{J}$ be a Banach space and X a $\mathbf{J}$-valued cadlag process. We call X a π-process if it is a $(\mathbf{L}, \mathbf{J}, \mathbf{K})$-primitive process for every Hilbert space $\mathbf{K}$ and $\mathbf{L} := \mathcal{L}(\mathbf{J}, \mathbf{K})$.

Remark 1 Since X^{u-} coincides with itself stopped at u, it follows from the property A.2 in Section 2.5 that for every process Y, integrable with respect to X^u in the sense of Section 2.3, the following equality holds:

$$\int Y\,dX^u = \int_{]0,\,u]} Y\,dX^u. \tag{2.9.1}$$

If $Y \in \mathcal{E}(\mathbf{L})$, the following expression $\int_{]0,\,u]} Y\,dX - Y_u(X_u - X_{u^-})$ has an immediate meaning and is clearly equal to the expression (2.9.1).

Remark 2 The second part of the above definition requires some comment. On the one hand, we have already mentioned in Section 2.6 that the so-defined class of processes is extremely wide. On the other hand, Exercise 1 gives a very simple process which is not a $(\mathbf{L}, \mathbf{J}, \mathbf{K})$-primitive process for every Banach space $\mathbf{K}$. This is the reason for restricting ourselves to the considerations of Hilbert spaces $\mathbf{K}$.

Remark 3 It immediately follows from the definition that if (u_n) is a localizing sequence for X and (v_n) is an increasing sequence of stopping times such that $\lim_n P\{v_n < t_m\} = 0$, the sequence $(u_n \wedge v_n)$ is also a localizing sequence for X. This is often used in the following situation: let Y be a locally bounded process (see Section 1.13), i.e., a process such that $1_{[0,\,v_n]} < Y$ is bounded for every n and some increasing sequence of stopping times v_n such that $\lim_n P\{v_n < t_m\} = 0$. Replacing u_n by $u_n \wedge v_n$, if necessary, we can therefore always assume that for such a localizing sequence (u_n), $1_{[0,\,u_n]}Y$ is bounded for all n.

As already mentioned in this chapter, we restrict ourselves to defining the stochastic integral for a class of processes which contains enough "basic processes" and proving that this class is stable for stochastic integration and change of variables (Ito formula; see Section 3). The fact that this class contains all "martingales," "local martingales," "semimartingales," etc., will be established in Section 8. We already know that Hilbert-valued cadlag square integrable martingales and Banach-valued processes with finite variation are π-processes.

2.10 Stochastic Integral with Respect to a π-Process

The stochastic integral with respect to a π-process X of every locally bounded predictable process Y (see Section 1.12) can be defined in the following way: we consider an increasing sequence $u(n)$ of stopping times localizing for X

$$P\Big[\bigcap_n \{u(n) < t\}\Big] = 0 \qquad \text{for every element } t \text{ of } T,$$

(see Section 2.9-Remark 3) such that $1_{[0,\,u(n)]}Y$ is bounded for every n. We can therefore define the process $(\int 1_{[0,\,u(n)]}Y\,dX^{u(n)})$ for every n.

We can prove the following assertion: let u and v be two stopping times such that X^u and X^v are L^2-primitive processes and $1_{[0,u]}Y$ and $1_{[0,v]}Y$ are bounded. Then, the processes $(\int 1_{[0,v]}Y\,dX^v)$ and $(\int 1_{[0,u]}Y\,dX^u)$ have the same restriction, up to P-equivalence, to $[0, u \wedge v[$.

In fact, the property is trivial for simple processes Y and therefore follows for a general predictable locally bounded process Y from the uniform approximation theorem of Section 2.5. The integral process Z will be then unambiguously defined by setting

$$Z = \left(\int Y\,dX^{u(n)}\right) \qquad \text{on} \quad [0, u(n)[.$$

In this general situation the cadlag integral process of Y with respect to X will still be denoted by $(\int Y\,dX)$.

The above construction immediately shows that the analog of the proposition of Section 2.4 holds:

Proposition *Let X be a* **J**-*valued π-process. Then for every Hilbert space* **K** *and for every locally bounded predictable* **L**-*valued process Y, the process $(\int Y\,dX)$ is a π-process.*

Remark We have already noted at the end of Section 1.9 that every process X whose paths are cadlag functions of finite variation is a π-process. The same argument as in Section 2.7 shows that the stochastic integral constructed above and the Stieltjes random integral define the same cadlag process up to P-equivalence.

2.11 Uniform Approximation and Dominated Convergence Theorems

Let X be a π-process and $(u(k))$ be a localizing sequence of stopping times. Let (Y_n) be a sequence of **L**-valued predictable processes which is uniformly bounded on each $[0, u(k)]$ and such that almost surely the sequence $Y_n(t, \omega)$ converges in **L** to $Y(t, \omega)$ for all $t \in T$.

Then, for every bounded random time v, the random variables

$$\int_{1_{[0,v]}} Y_n\,dX^{u(k)}, \; k \in \mathbf{N},$$

converge when $n \to \infty$ in $L^2(P)$ toward $\int 1_{[0,v]}Y\,dX^{u(k)}$. Since

$$\lim_{k\to\infty} P\{v < u(k)\} = 1$$

and

$$\int 1_{[0,v]}Y\,dX^{u(k)} = \int 1_{[0,v]}Y\,dX \qquad \text{on} \quad \{v < u(k)\},$$

we see immediately that the sequence $(\int 1_{[0,v]}Y_n\,dX)_{n\in\mathbf{N}}$ of random variables converges in probability toward $(\int 1_{[0,v]}Y\,dX)$.

We also remark that the theorems of Section 2.5 apply to the sequence (Y_n) and to $X^{u(k)}$ for every k.

All this can be summarized in the following statement:

Theorem *Let X, (Y_n) and $(u(k)_{k\in\mathbf{N}})$ be as just described.*

(1) *For every bounded stopping time v, the sequence of random $\mathbf{K}$-valued variables $(\int 1_{[0,v]} Y_n\, dX)_{n\in\mathbf{N}}$ converges when $n\to\infty$ to $(\int 1_{[0,v]} Y\, dX)$ in $L^0_{\mathbf{K}}(\Omega, \mathcal{F}, P)$.*

(2) *There exists moreover a subsequence $(n_r)_{r\in\mathbf{N}}$ of integers such that for P-almost all ω, the paths $Z_{n_r}(\omega)$ of the processes $Z_{n_r} := (\int Y_{n_r}\, dX)$ converge uniformly on every compact interval $[0,t]$ to the path $Z(\omega)$ of $Z := (\int Y\, dX)$.*

2.12 Stochastic Integral of Optional Processes with Respect to "Continuously Dominated" π-Processes

In the case of brownian motion, Poisson processes, etc., we have come across dominating measures of the following form:

$$\alpha(G) = E\left\{\int_T 1_G(s,\cdot)\, dA(s,\cdot)\right\}, \qquad G \text{ predictable},$$

where A is a positive increasing process. We will see later (in Section 10) that it is a rather general situation. In this case, the measure α is not only defined on predictable sets but also on $\mathcal{F}_{t_m} \otimes \mathcal{B}_{[0,t_m]}$. In particular, it is meaningful to speak of $\alpha([v])$, where v is a stopping time. Let us assume that the π-process X is dominated by measures $\alpha^{u\wedge u_n}$ with extensions giving measure zero to all graphs of stopping times (which is the case if A is continuous). Call $\alpha_n := \alpha^{u_n}$. Any function of the form $\sum \lambda_i 1_{]w_i, v_i]}$ is therefore equivalent for the measure α_n to the function $\sum \lambda_i 1_{[w_i, v_i[}$.

Since the σ-algebra $\mathcal{O}$ is generated by the stochastic intervals $[u, v[$, the predictable bounded processes constitute a dense family in $L^2_{\mathbf{L}}(\Omega', \mathcal{O}, \alpha_n)$. The same holds for $\mathcal{E}(\mathbf{L})$ which is dense in $L^2_{\mathbf{L}}(\Omega', \mathcal{P}, \alpha_n)$.

In other words, the natural imbedding $L^2_{\mathbf{L}}(\Omega', \mathcal{P}, \alpha_n) \to L^2_{\mathbf{L}}(\Omega', \mathcal{O}, \alpha_n)$ is an isomorphism of Hilbert spaces, and every locally bounded optional process Y is α_n-equivalent to a locally bounded predictable process Y'. The process $(\int Y\, dX)$ is therefore unambiguously defined as equal to $(\int Y'\, dX)$. It is, moreover, continuous according to Section 2.5 if X is itself continuous.

EXTENSIONS AND EXERCISES

1 For a real brownian motion β on $[0,1]$, there exists an increasing sequence (t_n) such that

$$E\left(\sum_n |\beta_{t_{n+1}} - \beta_{t_n}|\right) = +\infty.$$

Considering then the following Banach spaces $\mathbf{J} := \mathbf{R}$, $\mathbf{L} = \mathbf{K} := l^1(\mathbf{N})$ (the space of summable sequences of real members (x_n) with the norm $\sum |x_n|$) with the scalar multiplication $(x, \lambda) \rightsquigarrow \lambda x$ as a bilinear mapping from $(\mathbf{L}, \mathbf{J})$ into $\mathbf{K}$ and the following sequence of $\mathbf{L}$-valued simple functions,

$$Y_n = \sum_{k \leq n} 1_{]t_k, t_{k+1}]} e_k$$

where

$$e_k = (\delta_{ik})_{i \in \mathbf{N}}, \qquad \delta_{ij} = \begin{cases} 0 & \text{if } j \neq i \\ 1 & \text{if } i = j, \end{cases}$$

we can see that the brownian motion β has no dominating measure for the given bilinear mapping $\mathbf{L} \times \mathbf{J} \to \mathbf{K}$.

2 Let us consider the Banach spaces $\mathbf{J}, \mathbf{K}, \mathbf{L} \subset \mathcal{L}(\mathbf{J}; \mathbf{K})$ and $\mathbf{L}' \subset \mathcal{L}(\mathbf{K}; \mathbf{K}')$. We call $\tilde{\mathbf{L}}$ the closure in $\mathcal{L}(\mathbf{J}, \mathbf{K}')$ of the set of linear mappings $u \circ v$, $v \in \mathbf{L}$, $u \in \mathbf{L}'$. If X is an L^2-primitive process with respect to $(\mathbf{L}, \mathbf{J}, \mathbf{K})$ and $(\tilde{\mathbf{L}}, \mathbf{J}, \mathbf{K}')$ as well, show the analog of the proposition in Section 2.4.

3 N being a Poisson process, assumed to be cadlag, we denote by $T_1 < T_2 < \cdots < T_n \cdots$ the times of successive jumps of N. Describe the cadlag process X which is the solution of the following equation:

$$X_t = 1 + \int_{[0, t]} (X_{s^-})^2 \, dN_s.$$

Show that X is a π-process but is not an L^2-primitive process.

4 Let X be a π-process which is dominated by a process A such that for every ω, the Stieltjes measure $dA(\omega)$ is absolutely continuous with respect to the Lebesgue measure. In this case, extend the argument, given in Section 2.12 for optional processes, to the so-called progressively measurable processes which are the processes Y such that for all t, the restriction of Y to $\Omega \times [0, t]$ is $\mathcal{F}_t \otimes \mathcal{B}_{[0, t]}$ measurable.

5 Let $(\Omega, \mathcal{F}, P, (\mathcal{F}_t)_{t \in T})$ be a stochastic basis. We write $\mathbf{K}$ for the vector space of all real cadlag processes adapted to this stochastic basis. We suppose that there exists a positive mapping N defined on $\mathbf{K}$ such that.

(i) $N(X + Y) \leq N(X) + N(Y)$;
(ii) $N(\alpha X) = |\alpha| N(X)$ for each real number α;
(iii) $N(\sum_{n>0} X_n) \leq \sum_{n>0} N(X_n)$.

We call $\mathbf{H}$ the vector space of the elements X of $\mathbf{K}$ such that $N(X) < +\infty$. We suppose that for every element X of $\mathbf{H}$ and for every real $\mathcal{Q}$-simple process (cf. Section 2.2), the process Z defined by $Z_t = \int_{]0, t]} Y \, dX$ is such that

$$N(Z) \leq N(X) \sup_{t, \omega} |Y_t(\omega)|.$$

Prove with the same type of argument used in Section 2.5 that $\mathbf{H}$ is a complete space.

6 Let $(\Omega, \mathcal{F}, P, (\mathcal{F}_t)_{t \in T})$ be a stochastic basis. Let X be a real cadlag process adapted to this stochastic basis. We suppose that there exists a positive measure α such that for every real $\mathcal{Q}$-simple process (cf. Section 2.2), we have

$$E\left(\left|\int Y\,dX\right|\right) \leqslant \left(\int |Y|^2\,d\alpha\right)^{1/2}.$$

Is it possible to construct, as in Section 2.3, the stochastic integral $(\int Y\,dX)$ for all the processes Y which belong to $L^2(\Omega \times T, \mathcal{P}, \alpha)$?

HISTORICAL NOTES

Section 1. The content of this section is very classical. The fundamental role of the σ-algebra of predictable sets was disclosed by the Strasbourg school (see, for example, [Del], [Mey]; see also [Bur]). The basic properties of martingales and submartingales are due to Doob. What we call here the Doléans measure was introduced by Doléans to prove the uniqueness of the Meyer decomposition of a supermartingale (see [Dol-3]). The systematic use of the ring $\mathcal{Q}$, generated by the predictable rectangles, and of the additive Doléans function on $\mathcal{Q}$, associated with a process, has been promoted by the authors ([MeP-1] and [Met-6]). The same idea has been exploited by Föllmer ([Föl]).

Section 2. There are many books and lecture notes on stochastic integrals ([Sko], [Ito-4], [Gis-1], [McK], [Mey-4], [Kus], [McS-1], etc.) The prototype of the theories we give here is the Ito theory with respect to brownian motion. The idea of defining integrals with respect to square integrable martingales, already used in Doob's work [Doo-1], was exploited by Courreges [Cou] and Kunita and Watanabe [KuW]. The further extensions to real semimartingales are due mostly to the Strasbourg school. A masterly exposition of this theory is given in [Mey-4].

The consideration of integrals with respect to Hilbert-valued processes can be found first in [Kun]. The authors have developed this theory ([Met-5], [Pel-3], [MPi-2]). The construction given here, which rests on the domination property of π-processes (real or Hilbert-valued), is somewhat new.

Other points of view have been taken in the literature. The point of view of vector measures taken by the authors in several papers is considered in Sections 12 and 13 of this book.

Despite its interest (for control theory, for example), we do not cover here the "belated" stochastic integral, defined by MacShane, and the extension of the Stratonovic integral. For this theory the reader is referred to [McS-1]. The relations between Ito and Stratanovic integrals are illuminated in [Mey-4, Chapter VI] and in [Pro-4].

CHAPTER 2

THE ITO FORMULA

This chapter is devoted to the "formula of change of variables," the prototype of which is due to Ito [Ito-3]. The formula given here in the general context of Hilbert-valued π-processes is derived from successive contributions (see, in particular, [KuW], [Mey-4], [GrP]). The Ito formula is the basic formula of stochastic calculus, as is explained in the introduction to Section 3.

In Section 3 the formula is proved, and in Section 4, several applications are given.

For the reader's convenience, we have separately stated the formula in the finite-dimensional case (Section 3.3). But since the proof is essentially the same in both the finite- and infinite-dimensional Hilbert cases, we have given only one proof.

As we have already mentioned, the reader who cares only about real processes has simply to replace the tensor product $\otimes$ everywhere by the ordinary product of real numbers!

3 ITO FORMULA

3.1 Introduction

We take $T = R^+$. Let X and f be two real functions, X being defined on T and f being defined on the real line. Under the adequate hypothesis, we have

$$df(X) = f'(X)\,dX,$$

and this formula is fundamental to all calculations in differential equations. This formula can be more precisely written in the following form:

$$f(X_t) - f(X_0) = \int_{]0,\,t]} f'(X_s)\,dX_s.$$

Now let us consider the case in which X is a real continuous process, with f a real function defined on the real line; then, in general, we have not the previous equalities but, if X is continuous,

$$df(X) = f'(X)\,dX + \tfrac{1}{2} f''(X)\,d[X],$$

or more precisely,

$$f(X_t) - f(X_0) = \int_{]0,t]} f'(X_s)\,dX_s + \frac{1}{2}\int_{]0,t]} f''(X_s)\,d[X]_s,$$

where $[X]$ is an increasing process associated with X. This equality is called the *Ito formula*; it was proved for the first time for brownian motion in [Ito].

This formula plays a fundamental role in all the calculations involving differential stochastic equations.

Before proving this formula, we give the main idea of the proof.

Let us return to the above equality, given that X is a "regular" function: if $(t(k))_{1 \leqslant k \leqslant n}$ is an increasing sequence of times such that $t_1 = 0$ and $t_n = t$, we have

$$\begin{aligned} f(X_t) - f(X_0) &= \sum_{k=1}^{n-1} f(X_{t(k+1)}) - f(X_{t(k)}) \\ &= \sum_{k=1}^{n-1} f'(X_{t(k)})\left[X_{t(k+1)} - X_{t(k)}\right] + \sum_{k=1}^{n-1} R_k. \end{aligned}$$

Now, if $\sup_k [t(k+1) - t(k)]$ goes to zero, for "regular" functions f and X, the first sum converges to $\int_{]0,t]} f'(X_s)\,dX_s$ and the second sum converges to zero.

Now, if X is a process, in general, the second sum $\sum_{k=1}^{n-1} R_k$ does not go to zero. In this case, using the Taylor formula, we obtain

$$\begin{aligned} f(X_t) - f(X_0) &= \sum_{k=1}^{n-1} f'(X_{t(k)})\left[X_{t(k+1)} - X_{t(k)}\right] \\ &\quad + \frac{1}{2}\sum_{k=1}^{n-1} f''(X_{t(k)})\left[X_{t(k+1)} - X_{t(k)}\right]^2 + \sum_{k=1}^{n-1} R_k^*. \end{aligned}$$

For some functions f and for some processes X, when $\sup_k[t(k+1) - t(k)]$ goes to zero, the first sum converges to the stochastic integral $\int_{]0,t]} f'(X_{s^-})\,dX_s$, the second sum converges to $\frac{1}{2}\int_{]0,t]} f''(X_{s^-})\,d[X]_s$, and the third sum converges to zero.

Actually, for noncontinuous processes X it is convenient to consider stopping times $(u(k))_{k>0}$ instead of times $(t(k))_{k>0}$.

We shall prove the Ito formula for processes with values in a separable Hilbert space **H**. In our context, to suppose that **H** is separable is not a restriction; moreover, it is not more difficult to prove the Ito formula when

$\mathbf{H}$ is a Hilbert space than when $\mathbf{H}$ is a finite-dimensional vector space. It is also possible to prove this formula when $\mathbf{H}$ is a Banach space (cf. [GrP]).

In the sequel $(h_n)_{n>0}$ will be an orthogonal basis of $\mathbf{H}$. As we have done before, we shall consider a probabilized stochastic basis $(\Omega, \mathcal{F}, P, (\mathcal{F}_t)_{t \in \mathbf{R}_+})$ and suppose that this basis is complete and right continuous (cf. Section 1.1).

For the reader's convenience, we shall make the Ito formula explicit when $\mathbf{H}$ is finite dimensional in Section 3.3, before we state and prove it in Sections 3.7 and 3.8.

In the statement of the formula, processes with finite variation related to the π-process X come up, which we must now introduce.

3.2 Quadratic Variation

Let X be an $\mathbf{H}$-valued adapted process. It will be said to be with *finite quadratic variation* if for every $t \in T$, the following limit exists in $L^0_{\mathbf{R}}(\Omega, \mathcal{F}_t, P)$ (i.e., in probability):

$$D_t := \lim_{n \to \infty} \operatorname{prob} \sum_{k \geqslant 0} \|X_{(k+1)2^{-n} \wedge t} - X_{k2^{-n} \wedge t}\|^2 < \infty \text{ a.s.} \tag{3.2.1}$$

Among other things, the following proposition shows that every cadlag π-process has a finite quadratic variation and there exists a right continuous increasing version of the process D. This right continuous increasing version (defined up to P-equivalence) will be called *the quadratic variation of* X.

Proposition *If X is a cadlag Hilbert-valued π-process, it has finite quadratic variation. Moreover,*

(1) *the following equality holds for all $t \in T$,*

$$D_t = \|X_t\|^2 - \|X_0\|^2 - \int_{]0,t]} 2X_{s^-}\, dX_s \text{ a.s.}; \tag{3.2.2}$$

this proves, in particular, the existence of a cadlag modification for D.

(2) *Let, for every $n \in \mathbf{N}$, $(v(n,k) : k \in \mathbf{N})$ be an increasing sequence of bounded stopping times such that*

$$\lim_{n \to \infty} \sup_k (v(n, k+1) - v(n,k)) = 0 \text{ a.s.},$$

and for every t and n,

$$\lim_{k \to \infty} P\{v(n,k) < t\} = 0.$$

Then

$$D_t = \lim_{n \to \infty} \operatorname{prob} \sum_k \|X_{v(n,k+1) \wedge t} - X_{v(n,k) \wedge t}\|^2, \tag{3.2.3}$$

and there exists a subsequence (n_r) *of integers such that a.s. the paths of the processes* D^{n_r} *converge uniformly to the paths of* D *on every bounded interval if we set*

$$D_t^n := \sum_{k \geqslant 0} \|X_{v(n, k+1) \wedge t} - X_{v(n, k) \wedge t}\|^2. \tag{3.2.4}$$

(3) *For all* $t \in T$ *we have*

$$\sum_{s \leqslant t} \|X_s - X_{s^-}\|^2 < \infty \text{ a.s.} \tag{3.2.5}$$

(4) *The process* $(D_t - \sum_{s \leqslant t} \|X_s - X_{s^-}\|^2)_{t \in T}$ *is continuous.*

Proof We prove simultaneously the finiteness of the quadratic variation and the formula (3.2.2). We write

$$\begin{aligned} \|X_t\|^2 - \|X_0\|^2 &= \sum_{k \geqslant 0} \left[\|X_{(k+1)2^{-n} \wedge t}\|^2 - \|X_{k2^{-n} \wedge t}\|^2 \right] \\ &= \sum_{k \geqslant 0} \left[\|X_{(k+1)2^{-n} \wedge t} - X_{k2^{-n} \wedge t}\|^2 \right. \\ &\quad \left. + 2X_{k2^{-n} \wedge t}(X_{(k+1)2^{-n} \wedge t} - X_{k2^{-n} \wedge t}) \right]. \end{aligned}$$

But

$$\sum_{k \geqslant 0} \|X_{(k+1)2^{-n} \wedge t} - X_{k2^{-n} \wedge t}\|^2 = \|X_t\|^2 - \|X_0\|^2 - 2\int_{]0, t]} \Phi_n \, dX,$$

where we have set

$$\Phi_n := \sum_{k \geqslant 0} X_{k2^{-n}} 1_{]k2^{-n}, (k+1)2^{-n}]} \cdot$$

The sequence of simple processes Φ_n converges on Ω' toward the process $(X_{s^-} : s \in T)$. This process is clearly locally bounded, the sequence Φ_n is locally uniformly bounded, and (3.2.2) is an immediate consequence of Section 2.11. The same reasoning can be used to obtain (3.2.3). The existence of a subsequence which converges uniformly on every path follows from the uniform approximation Theorem 2.5.

Proof of (3) Assuming that we have extracted a subsequence with uniform convergence property for paths and denoting this sequence again by D^n, we have

$$\lim_n D_t^n(\omega) < \infty \qquad \text{for all} \quad \omega \text{ in a set of probability 1.}$$

Assuming that

$$\sum_{s \leqslant t} \|X_s(\omega) - X_{s^-}(\omega)\|^2 \geqslant 2d \qquad \text{for some} \quad d,$$

we immediately deduce from the right continuity of paths that $D_t^n(\omega) \geqslant d$

for n big enough. The equality

$$\sum_{s \leqslant t} \|X_s(\omega) - X_{s^-}(\omega)\|^2 = \infty$$

then contradicts the convergence of $D_t^n(\omega)$.

Proof of (4) We consider a particular family $v(n, k)$ defined in the following way: for every n,

$$\begin{aligned} v(n,0) &:= 0 \cdots \\ v(n,k+1) &:= \inf\{t : t > v(n,k), \|X_t - X_{v(n,k)}\| > 1/n\} \\ &\quad \wedge (v(n,k) + 1/n). \end{aligned}$$

Since the paths of X are cadlag, the family $v(n, k)$ meets the conditions in point (2) above. Extracting a subsequence (n_r) if necessary, we may assume that $\lim_n D_t^{n_r}(\omega)$ converges uniformly in t. For the extracted subsequence, which we still denote by $v(n, k)$, we have a fortiori the following inclusion for all n and ω:

$$\{v(n,k)(\omega) : k \in \mathbf{N}\} \supset \{s : \|X_s(\omega) - X_{s^-}(\omega)\| \geqslant 2/n\}. \qquad (3.2.6)$$

Let us then define

$$\begin{aligned} A(n,k) &:= \{\omega : \|X_{v(n,k)} - X_{v(n,k)^-}\| \geqslant 2/n\}, \\ T_n(\omega) &:= \{s : \|X_s(\omega) - X_{s^-}(\omega)\| < 2/n\}, \end{aligned}$$

and for $(\omega, t) \in]v(n, k-1), v(n, k)]$,

$$W_t^n(\omega) := \sum_{j=1}^{k} \|X_{v(n,j)}(\omega) - X_{v(n,j)^-}(\omega)\|^2 1_{A(n,j)}(\omega).$$

From the inclusion (3.2.6), we derive

$$\left| \sum_{s \leqslant t} \|X_s(\omega) - X_{s^-}(\omega)\|^2 - W_t^n(\omega) \right| \leqslant \sum_{\substack{s \leqslant t \\ s \in T_n(\omega)}} \|X_s(\omega) - X_{s^-}(\omega)\|^2.$$

From (3), for all bounded t_m, we have

$$\lim_{n \to \infty} \sum_{\substack{s \leqslant t_m \\ s \in T_n(\omega)}} \|X_s(\omega) - X_{s^-}(\omega)\|^2 = 0 \text{ a.s.}$$

The paths $t \rightsquigarrow (D_t^n(\omega) - W_t^n(\omega))$ converge uniformly to the corresponding path of the process $(D_t - \sum_{s \leqslant t} \|X_s - X_{s^-}\|^2)_{t \in T}$ uniformly on every bounded interval. Since the paths of $(D_t^n - W_t^n)_{t \in T}$ have no discontinuity greater than $1/n$, the limiting process is continuous. This completes the proof.

Notation The cadlag process of the quadratic variation of X will be denoted by $[X]$.

3.3 Ito formula: Finite-Dimensional Case

For the convenience of the reader who is not interested in the infinite-dimensional case, we state the Ito formula for a finite-dimensional **H**. But we give no particular proof, the present statement being a special case of the statement in Section 3.7, which will then be proved.

We suppose that **H** is a finite-dimensional vector space and $(h_j)_{1 \leqslant j \leqslant n}$ a basis of **H**. Let X be an **H**-valued cadlag π-process adapted to the complete stochastic basis $(\Omega, \mathfrak{F}, P, (\mathfrak{F}_t)_{t \in T})$, with $T = \mathbf{R}_+$.

We shall write

$$X = \sum_{j=1}^{n} X^j h_j.$$

We can define the following real processes with $1 \leqslant i \leqslant n$ and $1 \leqslant j \leqslant n$, the sum below being a.s. summable and the process $(X^i_{s^-})$ predictable, locally bounded, and therefore integrable with respect to X:

$$S_{i,j}(t) := \sum_{s \leqslant t} (X^i_s - X^i_{s^-})(X^j_s - X^j_{s^-}) = S_{j,i}(t),$$

$$Q(t) := \sum_{s \leqslant t} \left[f(X_s) - f(X_{s^-}) - \sum_{i=1}^{n} \frac{\partial f}{\partial x^i}(X_{s^-})(X^i_s - X^i_{s^-}) \right],$$

$$V_{i,j}(t) := X^i_t X^j_t - X^i_0 X^j_0 - \int_{]0,t]} (X^i_{s^-}\, dX^j_s + X^j_{s^-}\, dX^i_s),$$

$$C_{i,j}(t) := V_{i,j}(t) - S_{i,j}(t).$$

The processes $S_{i,j}$, Q, $V_{i,j}$, and $C_{i,j}$ are real cadlag processes of finite variation and $C_{i,j}$ is continuous; the processes $S_{i,i}$, $V_{i,i}$, and $C_{i,i}$ are increasing.

Proposition *We have then for every real function on* **H**, *which is twice continuously differentiable,*

$$\begin{aligned} f(X_t) - f(X_0) &= Q(t) + \int_{]0,t]} \sum_{i=1}^{n} \frac{\partial f}{\partial x^i}(X_{s^-})\, dX^i_s \\ &\quad - \frac{1}{2} \sum_{s \leqslant t} \sum_{i,j} \frac{\partial^2 f(X_{s^-})}{\partial x^i \partial x^j} (X^i_s - X^i_{s^-})(X^j_s - X^j_{s^-}) \\ &\quad + \frac{1}{2} \int_{]0,t]} \sum_{i,j} \frac{\partial^2 f(X_{s^-})}{\partial x^i \partial x^j}\, dV_{i,j}(s) \\ &= Q(t) + \int_{]0,t]} \sum_{i=1}^{n} \frac{\partial f(X_{s^-})}{\partial x^i}\, dX^i_s \\ &\quad + \frac{1}{2} \int_{]0,t]} \sum_{i,j} \frac{\partial^2 f(X_{s^-})}{\partial x^i \partial x^j}\, dC_{i,j}(s) \text{ a.s.} \end{aligned}$$

This formula is the same as that in Section 3.7: the matrix

$$\left[\left(X_s^i(\omega) - X_{s^-}^i(\omega)\right)\left(X_s^j(\omega) - X_{s^-}^j(\omega)\right)\right]_{i,j}$$

that occurs here is the expression in the given basis of the tensor product $(X_s(\omega) - X_{s^-}(\omega))^{\otimes_2}$; the bilinear mapping $(h, k) \rightsquigarrow \sum_{i,j}(\partial^2 f/\partial x^i \partial x^j)h^i k^j$ is the explicit expression of $f''(X_{s^-})(h \otimes k)$, $f''(X_{s^-})$ being considered as a linear form on $\mathbf{H} \otimes \mathbf{H}$ (see Section 3.5 below); the matrix processes V and C take their values in $\mathbf{H} \otimes \mathbf{H}$.

To help the reader, we recall a few simple facts about tensor products of Hilbert spaces.

3.4 Tensor Product and Hilbert–Schmidt Norm

This section may be disregarded by readers who are only interested in real-valued processes. In that case, in Section 3.5, 3.6 etc. they can replace the product $\otimes$ everywhere it occurs by the ordinary product of real numbers.

We shall write $\mathbf{H} \otimes \mathbf{H}$ for the tensor product of $\mathbf{H}$ by itself, denoting by $x \otimes y$ the tensor product of $x \in \mathbf{H}$ and $y \in \mathbf{H}$. If $x = y$, we shall write $x \otimes x = x^{\otimes_2}$

Let $(x_i, y_i)_{i \in I}$ be a finite family of pairs of elements of $\mathbf{H}$; let $z = \sum_{i \in I} x_i \otimes y_i$ be the element of $\mathbf{H} \otimes \mathbf{H}$ associated with this family. Similarly, we consider $z' = \sum_{j \in J} x'_j \otimes y'_j$.

If we put

$$\langle z, z' \rangle = \sum_{i \in I} \sum_{j \in J} \langle x_i, x'_j \rangle \langle y_i, y'_j \rangle,$$

this defines a scalar product on $\mathbf{H} \otimes \mathbf{H}$.

We shall denote by $\mathbf{H} \hat{\otimes}_2 \mathbf{H}$ the space $\mathbf{H} \otimes \mathbf{H}$ completed for the norm associated with this scalar product; the corresponding norm on $\mathbf{H} \hat{\otimes}_2 \mathbf{H}$ is called the Hilbert–Schmidt norm and will be denoted by $\|\cdot\|_2$. With this canonical extension of the scalar product defined above, $\mathbf{H} \hat{\otimes}_2 \mathbf{H}$ is a separable Hilbert space: more precisely, $(h_n \otimes h_m)_{n, m > 0}$ is a basis of $\mathbf{H} \hat{\otimes}_2 \mathbf{H}$. If x and y are two elements of $\mathbf{H}$, we have

$$\|x \otimes y\|_2 = \|x\|_{\mathbf{H}} \|y\|_{\mathbf{H}}.$$

The mapping $(x, y) \rightsquigarrow x \otimes y$ from $(\mathbf{H} \times \mathbf{H})$ into $(\mathbf{H} \hat{\otimes}_2 \mathbf{H})$ is a continuous bilinear mapping.

All the previous properties are well known and easy to prove. Let us recall that if $\mathbf{H}$ is d-dimensional, $\mathbf{H} \otimes \mathbf{H} = \mathbf{H} \hat{\otimes}_2 \mathbf{H}$ is isomorphic to the space of all $d \times d$ matrices: more precisely, let $(h_n)_{1 \leqslant n \leqslant d}$ be an orthonormal basis of $\mathbf{H}$, $(x_i, y_i)_{i \in I}$ a finite family of pairs of elements of $\mathbf{H}$, with $x_i = \sum_{n=1}^{d} x_{i,n} h_n$ and $y_i = \sum_{n=1}^{d} y_{i,n} h_n$, and (X^i, Y^i) the pairs of matrices defined by

$$X^i = \begin{bmatrix} x_{i,1} \\ \vdots \\ x_{i,d} \end{bmatrix} \quad \text{and} \quad Y^i = \begin{bmatrix} y_{i,1} \\ \vdots \\ y_{i,d} \end{bmatrix};$$

then the one-to-one mapping which associates the $d \times d$ matrix $((\sum_{i \in I} x_{i,j} y_{i,k}))_{j,k} = \sum_{i \in I} X^i (Y^i)^{\mathrm{tr}}$ to the element $(\sum_{i \in I} x_i \otimes y_i)$ of $\mathbf{H} \otimes \mathbf{H}$ is an isomorphism from $\mathbf{H} \otimes \mathbf{H}$ into the vector space of all $d \times d$ matrices.

3.5 Derivatives (Conventions)

Let $\mathbf{H}$ and $\mathbf{K}$ be two Hilbert spaces and f a $\mathbf{K}$-valued function defined on $\mathbf{H}$ and twice differentiable. Let f' and f'' denote the first and second derivative, respectively: the second derivative will be considered to be a $\mathbf{K}$-valued continuous bilinear mapping on $\mathbf{H} \times \mathbf{H}$ or a linear mapping from $\mathbf{H} \otimes \mathbf{H}$ into $\mathbf{K}$; if (x, y) is an element of $(\mathbf{H}, \mathbf{H} \otimes \mathbf{H})$, we shall write $[f''(x)](y)$ for the values of this second derivative considered at the point x and applied to the vector y.

We should notice (see [Tre]) that in the infinite-dimensional case, $f''(x)$ is continuous on $\mathbf{H} \otimes \mathbf{H}$ only for a stronger topology than the Hilbert–Schmidt one. For simplicity (see Remarks 3.9), we will make assumptions of continuity on f'' which are slightly stringent in the infinite-dimensional situation but are no restrictions in the finite-dimensional one.

3.6 Tensor Quadratic Variation. Processes S, Q, and $C^{\#}$

We consider a cadlag π-process X adapted to the complete stochastic basis $(\Omega, \mathcal{F}, P, (\mathcal{F}_t)_{t \in T})$ with values in the separable Hilbert space $\mathbf{H}$.

To this process X are associated the following processes of finite variation, which will enter the various forms of the Ito Formula.

Proposition *Let f be a $\mathbf{K}$-valued twice continuously differentiable function, defined on the Hilbert space $\mathbf{H}$, $\mathbf{K}$ being a Hilbert space. We suppose that the second derivative f'' of f is uniformly continuous as a mapping $\mathbf{H} \to \mathcal{L}(\mathbf{H} \hat{\otimes}_2 \mathbf{H}; \mathbf{K})$ on all the bounded subsets of $\mathbf{H}$. Then the following families $\{(X_s - X_{s^-})^{\otimes_2} : s \leqslant t\}$ and $\{f(X_s) - f(X_{s^-}) - f'(X_{s^-})(X_s - X_{s^-}) : s \leqslant t\}$ are a.s. summable in $\mathbf{H} \otimes_2 \mathbf{H}$ and $\mathbf{K}$, respectively.*

The process $(X_{s^-} : s \in T)$ is locally bounded and predictable. If we set

$$S_t := \sum_{s \leqslant t} (X_s - X_{s^-})^{\otimes_2}, \tag{3.6.1}$$

$$Q_t := \sum_{s \leqslant t} [f(X_s) - f(X_{s^-}) - f'(X_{s^-})(X_s - X_{s^-})], \tag{3.6.2}$$

$$V_t := X_t^{\otimes_2} - X_0^{\otimes_2} - \int_{]0,t]} (X_{s^-} \otimes dX_s + dX_s \otimes X_{s^-}), \tag{3.6.3}$$

$$C_t^{\#} := V_t - S_t, \tag{3.6.4}$$

the processes S, Q, V, and $C^{\#}$ are adapted cadlag processes of finite variation, with values in $\mathbf{H}\hat{\otimes}_2\mathbf{H}$, $\mathbf{K}$, $\mathbf{H}\hat{\otimes}_2\mathbf{H}$ *and* $\mathbf{H}\hat{\otimes}_2\mathbf{H}$, *respectively. Moreover, the process $C^{\#}$ is continuous.*

Proof The summability of the family $\{(X_s - X_{s^-})^{\otimes_2} : s \leqslant t\}$ follows from $\|(X_s - X_{s^-})^{\otimes_2}\|_2 \leqslant \|X_s - X_{s^-}\|_{\mathbf{H}}^2$ and from the proposition in Section 3.2(3).

Considering the process only on $[0, v^d[$ where $v^d := \inf\{t : \|X_t\| + [X_t] > d\}$, we reduce the situation to the case in which the quadratic variation of X is bounded by d. In other words, we consider the process X^{v^d} instead of X. It is clear that if the above theorem holds for the processes X^{v^d} for all d, it will be true for X.

From now on in this proof, we therefore assume that $[X]$ is uniformly bounded by d.

The Taylor formula gives

$$f(X_t) - f(X_{t^-}) - f'(X_{t^-})(X_t - X_{t^-})$$
$$= \left[\int_0^1 (1-s) f''[X_{t^-} + s(X_t - X_{t^-})]\, ds\right](X_t - X_{t^-})^{\otimes_2}.$$

Since for every $x \in \mathbf{H}$, the linear mapping $f''(x) \in \mathfrak{L}(\mathbf{H}\hat{\otimes}_2\mathbf{H}; \mathbf{K})$ is bounded by some constant C, as soon as $\|x\|_{\mathbf{H}}^2 \leqslant d$, we may write

$$\|f(X_s) - f(X_{s^-}) - f'(X_{s^-})(X_s - X_{s^-})\|_{\mathbf{K}} \leqslant C\|X_s - X_{s^-}\|_{\mathbf{H}}^2,$$

which ensures the summability a.s. of the family

$$\{f(X_s) - f(X_{s^-}) - f'(X_{s^-})(X_s - X_{s^-}) : s \leqslant t\}.$$

If we consider the elements h of $\mathbf{H}$ as operating linearly from $\mathbf{H}$ into $\mathbf{H}\hat{\otimes}_2\mathbf{H}$ through the formulas $k \rightsquigarrow h \otimes k$ or $k \rightsquigarrow k \otimes h$, the norm of this mapping for the uniform norm of operators being clearly $\|h\|$, the consistence of the definition (3.6.3) appears immediately.

Reproducing with slight modifications the proof in Section 3.2, and considering a family $\{v(n,k) : n, k \in \mathbf{N}\}$ of stopping times as in the proposition in Section 3.2, we may write for every n

$$X_t^{\otimes_2} - X_0^{\otimes_2} = \sum_{k \geqslant 0} \left[(X_{v(n,k+1)\wedge t})^{\otimes_2} - (X_{v(n,k)\wedge t})^{\otimes_2}\right]$$
$$= \sum_{k \geqslant 0} \Big[(X_{v(n,k+1)\wedge t} - X_{v(n,k)\wedge t})^{\otimes_2}$$
$$+ X_{v(n,k)\wedge t} \otimes (X_{v(n,k+1)\wedge t} - X_{v(n,k)\wedge t})$$
$$+ (X_{v(n,k+1)\wedge t} - X_{v(n,k)\wedge t}) \otimes X_{v(n,k)\wedge t}\Big],$$

and this leads to the formula

$$\sum_{k \geqslant 0} (X_{v(n,k+1)\wedge t} - X_{v(n,k)\wedge t})^{\otimes_2}$$
$$= X_t^{\otimes_2} - X_0^{\otimes_2} - \int_{]0,t]} \Phi^n \otimes dX - \int_{]0,t]} dX \otimes \Phi^n, \qquad (3.6.5)$$

where

$$\Phi^n := \sum_{k \geq 0} 1_{]v(n,k),\, v(n,k+1)]} X_{v(n,k)} \,. \tag{3.6.6}$$

The same argument used in Section 3.2 shows that for all t,

$$V_t = \lim_{n\to\infty} \operatorname{prob} V_t^n,$$

where

$$V_t^n = \sum_{k \geq 0} (X_{v(n,k+1)\wedge t} - X_{v(n,k)\wedge t})^{\otimes_2}. \tag{3.6.7}$$

Moreover, one can extract from the sequence V^n a subsequence V^{n_r}, the paths of which converge a.s. uniformly on every bounded interval to the paths of V.

The proof of the continuity of $C^{\#}$ goes therefore exactly as the proof of the continuity of C in Section 3.3. The fact that all the processes S, Q, V, and $C^{\#}$ are of finite variation follows immediately from their definition and the observation that their variations are majorized by the variations of the processes introduced in Section 3.2.

Therefore the proposition is proved, and we have, moreover, established the following corollary:

Corollary *Let for every* $n \in \mathbf{n}$, $(v(n,k) : k \in \mathbf{N})$ *be an increasing sequence of bounded stopping times with the properties*

$$\lim_{n\to\infty} \sup_k (v(n,k+1) - v(n,k)) = 0$$

and for every n and t,

$$\lim_{k\to 0} P\{v(n,k) < t\} = 0.$$

Then for all t, we have

$$X_t^{\otimes_2} - X_0^{\otimes_2} - \int_{]0,t]} (X_{s^-} \otimes dX_s + dX_s \otimes X_{s^-}) = \lim_{n\to\infty} \operatorname{prob} V_t^n,$$

where V_t^n *is defined by* (3.6.7).

It is, moreover, possible to extract a subsequence V^{n_r} *in such a way that a.s. the paths* $t \rightsquigarrow V^{n_r}(\omega)$ *of this subsequence converge uniformly on every compact interval* $[0, t_m]$.

Definition and notation The above cadlag $\mathbf{H} \hat{\otimes}_2 \mathbf{H}$ valued process V will be called the *tensor quadratic variation of* X and denoted by $[X]$.

3.7 The Ito Formula

We may now state the Ito formula[1] in the form of a theorem.

[1] In the general form, where we state it, for possibly discontinuous processes, the formula could be perhaps more properly called Ito–Watanabe–Kunita–Meyer formula.

Theorem *Let X be a cadlag* **H**-*valued* π-*process, where* **H** *is a Hilbert space. We consider the associated processes* $[X]$, Q, *and* $C^{\#}$ *with finite variation, as defined in Section* 3.6.

Let **K** *be a Hilbert space and f a* **K**-*valued twice continuously differentiable function on* **H**. *We assume that the second derivative f'' of f is uniformly continuous as a mapping* $\mathbf{H} \to \mathcal{L}(\mathbf{H} \hat{\otimes}_2 \mathbf{H}; \mathbf{K})$ *on all the bounded subsets of* **H**.

Then the following three cadlag processes

$$(f(X_t) - f(X_0))_{t \in T},$$

$$\left(Q_t - \frac{1}{2} \sum_{s \leq t} f''(X_s)(X_s - X_{s^-})^{\otimes_2} + \int_{]0,\, t]} f'(X_{s^-})\, dX_s + \frac{1}{2} \int_{]0,\, t]} f''(X_{s^-}) d[X]_s \right)_{t \in T},$$

$$\left(Q_t + \int_{]0,\, t]} f'(X_{s^-})\, dX_s + \frac{1}{2} \int_{]0,\, t]} f''(X_{s^-})\, dC_s^{\#} \right)_{t \in T}$$

are equal up to P-equivalence.

Consequence If X is a cadlag π-process, the process $f(X)$, where f is defined as in the above theorem, is also a cadlag π-process.

3.8 Proof of the Ito Formula

Let $(b(n))_{n \in \mathbf{N}}$ be a decreasing sequence of real strictly positive numbers such that $b(n) \leq 1/n^2$ for all n.

First, we construct a family $\{(v(n,k) : n, k \in \mathbf{N}\}$ of stopping times with the following properties:

(i) for every n, $\{v(n,k) : k \in \mathbf{N}\}$ is an increasing family of stopping times such that for all $t \in T$ $\lim_{k\to\infty} P\{v(n,k) < t\} = 0$;

(ii) $\lim_{n\to\infty} \sup_k (v(n,k+1) - v(n,k)) = 0$;

(iii) for all n and k, $\|X_{v(n,\,k+1)^-} - X_{v(n,\,k)}\| \leq b(n)$ on $\{v(n,k+1) < \infty\}$;

(iv) the finite sums

$$V_t^n = \sum_{k \geq 0} (X_{v(n,\,k+1) \wedge t} - X_{v(n,\,k) \wedge t})^{\otimes_2}$$

and

$$D_t^n = \sum_{k \geq 0} \|X_{v(n,\,k+1) \wedge t} - X_{v(n,\,k) \wedge t}\|^2,$$

define processes, the paths of which a.s. converge uniformly to the paths of the processes $[X]$ and $[X]$, respectively, on any compact interval.

To show the existence of such a family $\{v(n,k)\}$, we first define recursively the family $\{v'(n,k) : n, k \in \mathbf{N}\}$ by

$$v'(n,k) := 0;$$

$$v'(n,k+1) := (v(n,k) + 1/n) \wedge \inf\{s : s \in T, s > v(n,k), \|X_s - X_{v(n,k)}\| > b(n)\}$$

(with the usual convention that $\inf \emptyset = +\infty$). Applying the propositions in Sections 3.2 and 3.6, we can extract a subfamily, which we call $v(n,k)$, possessing all the required properties.

As in the proof of Section 3.6, by considering the processes stopped strictly before some proper stopping times v, we may reduce the proof to the case in which the processes X, $[X]$, Q, and $C^{\#}$ are uniformly bounded in norm by some constant a.

We then have to prove under this assumption that for all t the three processes of the theorem are a.s. equal. Since t is now fixed throughout the proof, we write $v(n,k)$ instead of $v(n,k) \wedge t$ for simplicity. We define

$$A(n,k) := \{\omega : \|X_{v(n,k)} - X_{v(n,k)^-}\| > 1/n\},$$

$$B(n,k) := \Omega - A(n,k).$$

For any integer n we have

$$f(X_t) - f(X_0) = \sum_{k \geq 0} \left[f(X_{v(n,k+1)}) - f(X_{v(n,k)}) \right] \left[1_{A(n,k+1)} + 1_{B(n,k+1)} \right]. \tag{3.8.1}$$

Let us also define the random variable $R_{n,k}$ by

$$\begin{aligned} f(X_{v(n,k+1)}) - f(X_{v(n,k)}) &= f'(X_{v(n,k)})\left[X_{v(n,k+1)} - X_{v(n,k)}\right] \\ &\quad + \tfrac{1}{2} f''(X_{v(n,k)})(X_{v(n,k+1)} - X_{v(n,k)})^{\otimes 2} \\ &\quad + R_{n,k}(\omega). \end{aligned} \tag{3.8.2}$$

This Taylor formula will actually be used only on the set $B(n,k+1)$.

We write

$$f(X_t) - f(X_0) = \sum_{k \geq 0} \sum_{i=1}^{5} a_{n,k}^i$$

with

$$a_{n,k}^1 = f'(X_{v(n,k)})(X_{v(n,k+1)} - X_{v(n,k)});$$

$$a_{n,k}^2 = -\tfrac{1}{2} f''(X_{v(n,k)})(X_{v(n,k+1)} - X_{v(n,k)})^{\otimes 2} 1_{A(n,k+1)};$$

$$a_{n,k}^3 = R_{n,k} 1_{B(n,k+1)};$$

$$\begin{aligned} a_{n,k}^4 = \big[&-f'(X_{v(n,k)})(X_{v(n,k+1)} - X_{v(n,k)}) \\ &+ f(X_{v(n,k+1)}) - f(X_{v(n,k)}) \big] 1_{A(n,k+1)}; \end{aligned}$$

$$a_{n,k}^5 = \tfrac{1}{2} f''(X_{v(n,k)})(X_{v(n,k+1)} - X_{v(n,k)})^{\otimes 2}.$$

We now study the limit of $\sum_{k\geqslant 0} a^i_{n,k}$, $i = 1, \ldots, 5$, when n tends to infinity.

(1) Limit of $\sum_{k\geqslant 0} a^1_{n,k}$: If we define

$$Z^n := \sum_{k\geqslant 0} f'(X_{v(n,k)}) 1_{]v(n,k),\, v(n,k+1)]},$$

we may write

$$\sum_{k\geqslant 0} a^1_{n,k} = \int_{]0,t]} Z^n_s \, dX_s.$$

As an immediate consequence of Theorem 2.11, we derive

$$\lim_{n\to\infty} \text{prob} \sum_{k\geqslant 0} a^1_{n,k} = \int_{]0,t]} Z_{s^-} \, dX_s, \qquad \text{where} \quad Z := f'(X).$$

(2) Limit of $\sum_{k\geqslant 0} a^2_{n,k}$: Let us introduce the process S^n

$$S^n_s(\omega) := \sum_{k\geqslant 0} \Bigg(\sum_{j\leqslant k} (X_{v(n,j)}(\omega) - X_{v(n,j)^-}(\omega))^{\otimes_2} 1_{A(n,j)}(\omega) \Bigg) 1_{]v(n,k-1),\, v(n,k)]}(s,\omega).$$

As in the proof of Section 3.2 concerning the processes W^n, we see that the processes S^n converge a.s. uniformly by path to the process

$$S_s := \sum_{r\leqslant s} (X_r - X_{r^-})^{\otimes_2} \qquad \text{on } [0,t].$$

Since the total variation of the path $S(\omega) - S^n(\omega)$ on $[0,t]$ is clearly smaller than

$$\sum_{\substack{s\leqslant t \\ s\in T_n(\omega)}} \|X_s(\omega) - X_{s^-}(\omega)\|^2, \qquad T_n(\omega) = \{s : \|X_s(\omega) - X_{s^-}(\omega)\| \leqslant 1/n\},$$

we see that the total variation of the process $S - S_n$ goes to zero a.s. If we now define the process W^n by

$$W^n_s(\omega) := \sum_{k\geqslant 0} \Bigg(\sum_{j=1}^{k} (X_{v(n,j)}(\omega) - X_{v(n,j-1)}(\omega))^{\otimes_2} 1_{A(n,j)} \Bigg) \times 1_{]v(n,k-1),\, v(n,k)]}(s,\omega),$$

and notice that on $A(n,j)$, we have

$$\|X_{v(n,j-1)} - X_{v(n,j)^-}\| \leqslant (1/n)^2$$

and

$$\|X_{v(n,j)} - X_{v(n,j)^-}\| \geqslant (1/n),$$

then, for all $s \leqslant t$ the total variation of $W^n - S^n$ is smaller than

$$\frac{2}{n} \sum_{j>0} \|X_{v(n,j)} - X_{v(n,j)^-}\|^2 1_{A(n,j)},$$

and this shows that the sequence of total variations of processes $(W^n - S^n)$ converges to zero a.s. uniformly on $[0, t]$. We may write

$$\sum_{k \geqslant 0} a_{n,k}^2 = -\frac{1}{2}\int_{]0,t]} U^n \, dW^n = -\frac{1}{2}\int_{]0,t]} U^n \, dS^n + \frac{1}{2}\int_{]0,t]} U^n \, d(S^n - W^n),$$

where

$$U^n = \sum_{k \geqslant 0} f''(X_{v(n,k)}) 1_{]v(n,k), v(n,k+1)]}.$$

But U^n is bounded by some constant C and therefore

$$\lim_{n \to \infty} \int_{]0,t]} U^n \, d(S^n - W^n) = 0 \text{ a.s.}$$

Using the fact that the total variation of $S - S^n$ on $[0, t]$ goes to zero a.s. and that $U_s^n(\omega)$ converges boundedly to $f''(X_{s^-}(\omega))$ for all s and ω, we immediately obtain

$$\lim_{n \to \infty} \sum_{k \geqslant 0} a_{n,k}^2 = -\frac{1}{2}\int_{]0,t]} f''(X_{s^-}) \, dS_s.$$

(3) Limit of $\sum_{k \geqslant 0} a_{n,k}^3$: The Taylor formula and the definition of $R_{n,k} 1_{B(n,k+1)}$ show that

$$1_{B(n,k+1)} \|R_{n,k}(\omega)\| \leqslant r_n \sup_k \|X_{v(n,k+1)} - X_{v(n,k)}\|^2 1_{B(n,k+1)}$$

where r_n goes to zero when r goes to infinity. Therefore,

$$\lim_{n \to \infty} \sum_{k \geqslant 0} a_{n,k}^3 = 0 \text{ a.s.}$$

(4) Limit of $\sum_{k \geqslant 0} a_{n,k}^4$: We write

$$\sum_{k \geqslant 0} a_{n,k}^4 = \sum_{k \geqslant 0} -\big[f'(X_{v(n,k)})(X_{v(n,k+1)} - X_{v(n,k+1)^-})$$
$$+ f(X_{v(n,k+1)}) - f(X_{v(n,k+1)^-})\big] 1_{A(n,k+1)} + O(1/n),$$

where

$$O\left(\frac{1}{n}\right) = \sum_{k \geqslant 0} -\big[f'(X_{v(n,k)})(X_{v(n,k+1)^-} - X_{v(n,k)})$$
$$+ f(X_{v(n,k+1)^-}) - f(X_{v(n,k)})\big] 1_{A(n,k+1)}.$$

Using the Taylor formula, we obtain

$$\left\| O\left(\frac{1}{n}\right) \right\| \leqslant C \sum_{k \geqslant 0} \|X_{v(n,k+1)^-} - X_{v(n,k)}\|^2 1_{A(n,k+1)},$$

which gives, by the same argument in point (2) above,

$$\|O(1/n)\| \leqslant 2(C/n) S_t.$$

Therefore,

$$\lim_{n\to\infty} \sum_{k\geqslant 0} a^4_{n,k} = Q_t \text{ a.s.}$$

(5) Limit of $\sum_{k\geqslant 0} a^5_{n,k}$: We may write

$$a^5_{n,k} = \tfrac{1}{2} f''(X_{v(n,k)})\{X^{\otimes_2}_{v(n,k+1)} - X^{\otimes_2}_{v(n,k)}$$
$$- X_{v(n,k)} \otimes (X_{v(n,k+1)} - X_{v(n,k)}) - (X_{v(n,k+1)} - X_{v(n,k)}) \otimes X_{v(n,k)}\}.$$

If we define X^n as

$$X^n := \sum_{k\geqslant 0} X_{v(n,k)} 1_{]v(n,k),\, v(n,k+1)]},$$

we obtain the following equality:

$$\sum_{k\geqslant 0} a^5_{n,k} = \int_{]0,t]} \frac{1}{2} f''(X^n_s)\, dX^{\otimes_2}_s$$
$$- \frac{1}{2}\int_{]0,t]} f''(X^n_s)(X^n_s \otimes dX_s) - \frac{1}{2}\int_{]0,t]} f''(X^n_s)(dX_s \otimes X^n_s)$$

(we note that $X^{\otimes_2}_s$ is a π-process in view of the proposition in Section 3.6, formula (3.6.3)). If we consider $f''(X^n_s)(X^n_s \otimes \cdot)$ to be a process taking its values in $\mathcal{L}(\mathbf{H};\mathbf{K})$ and the same for the process $f''(X^n_s)(\cdot \otimes X^n_s)$, we derive immediately from formula 3.6.3 (where V stands for $[X]$) and 2.10,

$$\int_{]0,t]} f''(X^n_s)(X^n_s \otimes dX_s) + \int_{]0,t]} f''(X^n_s)(dX_s \otimes X^n_s)$$
$$\underset{\text{prob}}{\overset{n\to\infty}{\to}} \int_{]0,t]} f''(X_{s-})\, dX^{\otimes_2}_s - \int_{]0,t]} f''(X_{s-})\, d[X]_s.$$

Therefore, finally,

$$\operatorname*{lim\,prob}_{n\to\infty} \sum_{k\geqslant 0} a^5_{n,k} = \frac{1}{2}\int_{]0,t]} f''(X_{s-})\, d[X]_s.$$

In conclusion, we have proved that

$$f(X_t) - f(X) = Q_t - \frac{1}{2}\sum_{s\geqslant t} f''(X_{s-})(X_s - X_{s-})^{\otimes_2}$$
$$+ \int_{]0,t]} f'(X_{s-})\, dX_s + \frac{1}{2}\int_{]0,t]} f''(X_{s-})\, d[X]_s \text{ a.s.}$$

This shows the equality, up to P-equivalence, of the two first cadlag processes of the theorem. The equality of the second and the third processes immediately follows from the definition of

$$C^{\#}_t := [X]_t - \sum_{s\leqslant t} (X_s - X_{s-})^{\otimes_2}.$$

3.9 Remarks

(1) In Section 3.5, we supposed that f'' was uniformly continuous on all the bounded subsets of $\mathbf{H}$. Actually, in the proof of Section 3.7, we used exactly the following property:

for each pair (a, ϵ) of positive numbers, there exists a positive number η such that $\|x\| \leqslant a$ and $\|y\| \leqslant \eta$ imply $\|f(x+y) - f(x) - f'(x)y - \frac{1}{2}f''(x)y^{\otimes_2}\|_{\mathbf{K}} \leqslant \epsilon\|y\|^2_{\mathbf{H}}$.

(2) As already noted (see Section 3.5), for a twice differentiable function, the derivatives $f''(x)$ are in general elements of $\mathcal{L}(\mathbf{H}\hat{\otimes}_1\mathbf{H}; \mathbf{K})$, where $\mathbf{H}\hat{\otimes}_1\mathbf{H}$ is the completion of $\mathbf{H}\otimes\mathbf{H}$ for a stronger topology than the Hilbert–Schmidt one (see [Tre]). Since the norm $\|\cdot\|_1$ in $\mathbf{H}\hat{\otimes}_1\mathbf{H}$ is such that $\|x^{\otimes_2}\|_1 = \|x\|^2_{\mathbf{H}}$, the argument in Section 3.2 shows in fact that S, V, and C take their values in $\mathbf{H}\hat{\otimes}_1\mathbf{H}$. The theorem in Section 3.5 remains valid if f'' is only assumed to be uniformly continuous from H into $\mathcal{L}(\mathbf{H}\hat{\otimes}_1\mathbf{H}; \mathbf{K})$ (see [Met-5] and [GrP]). In the latter paper, the case of Banach-valued processes is considered.

4 APPLICATIONS OF THE ITO FORMULA

We begin this section by examining some consequences of the existence of quadratic variation and then consider two particular applications of the Ito formula which are illustrative of the stochastic calculus.

MUTUAL VARIATIONS OF TWO π-PROCESSES

4.1 The Processes $[X, Y]$ and $[X, Y]$

Let $\mathbf{H}$ be a Hilbert space with scalar product written xy to simplify. Since the process (X_{s^-}) is predictable locally bounded as soon as X is cadlag adapted, the stochastic integral process $(\int_{]0,t]} X_{s^-}\, dY_s)$ exists for every cadlag adapted process X and every π-process Y. We may therefore define for two $\mathbf{H}$-valued cadlag π-processes:

$$[X, Y]_t := X_t Y_t - X_0 Y_0 - \int_{]0,t]} (X_{s^-}\, dY_s + Y_{s^-}\, dX_s), \qquad (4.1.1)$$

and denote the right continuous version of this process by $[X, Y]$.

We remark that for $Y = X$, the process $[X, X]$ is nothing other than the process $[X]$, the quadratic variation of X as defined in Section 3.2.

By exactly replicating the argument of Section 3.2, it is easy to show that for every family of stopping times $\{v(n,k) : n, k \in \mathbf{N}\}$ such that

(a) for every n, the sequence $(v(n,k): k \in \mathbf{N})$ is increasing and $\lim_{n\to\infty} \sup_k (v(n,k+1) - v(n,k)) = 0$ a.s.;
(b) for every $t \in T$, $\lim_{k\to\infty} P\{v(n,k) < t\} = 0$,

we have

$$[X, Y]_t = \lim_{n\to\infty} \text{prob}\, D_t^n(X, Y), \tag{4.1.2}$$

where

$$D_t^n(X, Y) := \sum_{k \geqslant 0} (X_{v(n,k+1)\wedge t} - X_{v(n,k)\wedge t})(Y_{v(n,k+1)\wedge t} - Y_{v(n,k)\wedge t}).$$

Moreover, there exists a subsequence $(D_t^{n_r})_{r\in\mathbf{N}}$ which converges uniformly on every bounded interval almost surely.

From the elementary formula

$$xy = \tfrac{1}{4}\left[\|x+y\|^2 - \|x-y\|^2\right],$$

it can easily be deduced that

$$[X, Y] = \tfrac{1}{4}([X+Y] - [X-Y]). \tag{4.1.3}$$

Noticing now that for every $h \in \mathbf{H}$ the mappings $x \curvearrowright x \otimes h$ and $x \curvearrowright h \otimes x$ are continuous linear mappings from $\mathbf{H}$ into $\mathbf{H} \hat{\otimes}_2 \mathbf{H}$ with norm smaller than $\|h\|$, the following process $[X, Y]$ can also be defined:

$$[X, Y]_t := X_t \otimes Y_t - X_0 \otimes Y_0 - \int_{]0,t]} (X_{s^-} \otimes dY_s + dX_s \otimes Y_{s^-}). \tag{4.1.4}$$

With the same family $\{v(n,k)\}$ of stopping times as above, we can write

$$\tilde{D}_t^n(X, y) := \sum_{k \geqslant 0} (X_{v(n,k+1)\wedge t} - X_{v(n,k)\wedge t}) \otimes (Y_{v(n,k+1)\wedge t} - Y_{v(n,k)\wedge t}),$$

and we have

$$[X, Y]_t = \lim_{n\to\infty} \text{prob}\, \tilde{D}_t^n(X, Y),$$

and for some subsequence $(n_r)_{r\in\mathbf{N}}$, the uniform convergence of $\tilde{D}_t^n(X, Y)$ for almost all paths holds on any bounded interval.

By comparing this with the definitions in Section 3.6, we notice immediately that the process $[X, X]$ just defined is nothing else but the process $[X]$ of Section 3.6.

Definition The processes $[X, Y]$ (resp. $[X, Y]$) is called the *mutual variation* of X and Y (resp. the *tensor mutual variation* of X and Y).

From their definition (see also (4.1.3), the processes $[X, Y]$ and $[X, Y]$, with values in $\mathbf{R}$ and $\mathbf{H} \hat{\otimes}_2 \mathbf{H}$, respectively, are with finite variation.

Remark and Extension: If the reader knows about the tensor product of two Hilbert spaces **H** and **K**, he will immediately notice that formula (4.1.4) extends readily to the case in which processes X and Y are **H** and **K**-valued, respectively. Then, $[X, Y]$ is still called the tensor mutual variation of X and Y. In the particular case $\mathbf{K} = \mathbf{R}$, $\mathbf{H} \otimes \mathbf{R}$ is isomorphic to **H** if $h \otimes \lambda$ is identified with λh. The definition of $[X, Y]$ as an **H**-valued process in this case can be rephrased directly by the reader.

4.2 Stochastic Differential Calculus

Proposition *Let* **H**, **J**, *and* **K** *be three Hilbert spaces. We assume that X is an* **H**-*valued cadlag adapted π-process and Y (resp. Z) is an $\mathfrak{L}(\mathbf{H}, \mathbf{J})$- (resp. $\mathfrak{L}(\mathbf{J}, \mathbf{K})$-) valued strongly predictable and uniformly bounded process. We denote by U and V the processes defined by*

$$U_t = \int_{]0,t]} Y\, dX \qquad \text{and} \qquad V_t = \int_{]0,t]} ZY\, dX.$$

Then, (1) *U and V are π-processes and*

$$V_t = \int_{]0,t]} Z\, dU$$

(i.e., with the symbolic differential notation, if $dU = Y\, dX$ and $dV = ZY\, dX$, then $dV = Z\, dU$).

(2) *If X is a martingale, then U and V are also martingales.*

(3) *We have*

$$[U, X]_t - [U, X]_0 = \int_{]0,t]} Y \otimes I\, d[X, X] \qquad I \text{ being the identity mapping}$$

and

$$[U, U]_t - [U, U]_0 = \int_{]0,t]} Y^{\otimes_2} d[X, X]$$

(i.e., with the symbolic differential notation, $d[U, X] = Y\, d[X, X]$ and $d[U, U] = Y^{\otimes_2} d[X, X]$).

(4) *We have also*

$$\langle U, X \rangle_t = \mathrm{Tr}[U, X] \leqslant \int_{]0,t]} \|Y\|\, d\langle X, X \rangle,$$

$$\langle U, U \rangle_t = \mathrm{Tr}[U, U] \leqslant \int_{]0,t]} \|Y\|^2\, d\langle X, X \rangle.$$

Proof We have already noted in Section 2.10 that U and V are π-processes. Moreover, all the properties (1), (2), and (3) are easily checked when $Y = y \cdot 1_A$ and $Z = z 1_B$ with $y \in \mathfrak{L}(\mathbf{H}, J)$, $z \in \mathfrak{L}(J, \mathbf{K})$, $A \in \mathfrak{R}$, and $B \in \mathfrak{R}$ (cf. Section 1.7). Then, by linearity, these same properties are true when Y and Z are $\mathcal{Q}$-simple processes, and they follow easily, in the general case, from the uniform approximation the theorem of Section 2.5.

4.3 Formula $[X, V] = 0$ for V with Finite Variation

Proposition *Let X, Y, and V be three cadlag adapted π-processes with values in the Hilbert space* **H**. *If X and Y are continuous and if V is with bounded variation, we have*

$$[V, Y] = 0, \qquad [X + V, Y] = [X, Y], \qquad [X + V, X + V] = [X, X],$$

and the same relations hold for the tensor mutual variation.

Proof By linearity, the second and the third equalities are consequences of the first one. Let us prove that $[V, Y] = 0$. Let ϵ be a positive number and $(u^{\epsilon}(k))_{k>0}$ be the sequence of stopping times defined recursively by

$$u^{\epsilon}(1) = 0$$

$$u^{\epsilon}(k+1) = \inf\{t : \|Y_t - Y_{u(k)}\| + \|V_t - V_{u(k)}\| > \epsilon\}.$$

For every $t \in T = [0, t_m]$, we have

$$\begin{aligned} &\left\| \sum_{k>0} \left(V_{u^{\epsilon}(k+1)\wedge t} - V_{u^{\epsilon}(k)\wedge t}\right) \otimes \left(Y_{u^{\epsilon}(k+1)\wedge t} - Y_{u^{\epsilon}(k)\wedge t}\right) \right\| \\ &\qquad \leqslant \epsilon \sum_{k>0} \| V_{u^{\epsilon}(k+1)\wedge t_m} - V_{u^{\epsilon}(k)\wedge t_m} \|. \end{aligned} \tag{4.3.1}$$

But the quantity on the right side of the last inequality goes to zero when $\epsilon\downarrow 0$. Extracting a subsequence (ϵ_r) for which the expressions in (4.3.1) converge a.s. (see Section 4.1) to $[V, Y]_t$ and $\lim \epsilon_r |V| = 0$, respectively, we obtain $[V, Y]_t = 0$. Since $[V, Y]$ is cadlag, it is clearly P-equivalent to zero.

4.4 Quadratic Variation of Martingales

Let M be a cadlag square integrable martingale, with values in the Hilbert space **H**. For every n, we define the stopping time

$$v_n := \inf\{t : \|M_t\| > n\}.$$

The process $(M_{t^-})_{t\in[0, t_m]}$ is therefore bounded on $[0, v_n]$. As a consequence (see the proposition in Section 2.6), the process $(\int M_{s^-}\, dM_s)$ stopped at v_n is a square integrable martingale. Then, if we write again the definition of $[M]$, we see that the process $[M] - \|M\|^2 + \|M_0\|^2$, stopped at v_n, is a square integrable martingale. Hence follows the equality on $]0, v_n]$ of the Doléans measure of $[M]^{v_n}$ and $\|M\|^{2v_n}$. Taking the limit when $n\uparrow\infty$, we see that $E([M]_{t_m}) < \infty$ and the processes $[M]$ and $\|M\|^2$ have the same Doléans measure on $\Omega \times]0, t_m]$.

As a consequence, we see that the dominating measure α introduced in Section 2.6 for a square integrable martingale M and which is nothing else but the Doléans measure of $\|M\|^2$ can also be written

$$\alpha(G) = E\Big(\int_{]0,t_m]} 1_G \, d[M]\Big) \qquad \text{for every} \quad G \in \mathcal{P}. \tag{4.4.1}$$

The same argument, leading to the observation that $[M] - \|M\|^2$ is a martingale, shows that $[\![M]\!] - M^{\otimes_2}$ is an $\mathbf{H} \hat{\otimes}_2 \mathbf{H}$-valued martingale as soon as M is an $\mathbf{H}$-valued square integrable martingale.

We summarize all this in the following proposition.

Proposition *Let M be a cadlag* $\mathbf{H}$*-valued square integrable martingale on $T := [0, t_m]$. Then, the processes $[M] - \|M\|^2$ and $[\![M]\!] - M^{\otimes_2}$ are martingales. In particular, we have for every $s < t$, $s, t \in T$,*

$$E\{(M_t - M_s)^{\otimes_2} \mid \mathcal{F}_s\} = E\{(M_t^{\otimes_2} - M_s^{\otimes_2}) \mid \mathcal{F}_s\}$$
$$= E\{([\![M, M]\!]_t - [\![M, M]\!]_s) \mid \mathcal{F}_s\}$$

and

$$E\{\|M_t - M_s\|^2 \mid \mathcal{F}_s\} = E\{\|M_t\|^2 - \|M_s\|^2 \mid \mathcal{F}_s\} = E\{[M]_t - [M]_s \mid \mathcal{F}_s\}.$$

Proof Everything in this proposition has been proved above, except for the first equality in the two stated formulas. But this follows immediately from

$$(M_t - M_s)^{\otimes_2} = M_t^{\otimes_2} - M_s^{\otimes_2} - M_s \otimes (M_t - M_s) - (M_t - M_s) \otimes M_s,$$
$$\|M_t - M_s\|^2 = \|M_t\|^2 - \|M_s\|^2 - 2M_s(M_t - M_s),$$

and from

$$E\{M_s \otimes (M_t - M_s) \mid \mathcal{F}_s\} = E\{(M_t - M_s) \otimes M_s \mid \mathcal{F}_s\} = 0,$$
$$E\{M_s(M_t - M_s) \mid \mathcal{F}_s\} = 0.$$

as a result of the martingale property.

4.5 An Expression of the Domination Property for Some π-Processes

Let M be a cadlag $\mathbf{H}$-valued square integrable martingale. Standard measure theoretic arguments show that the formula (4.4.1) implies that

$$\int Y \, d\alpha = E\Big(\int_{[0,t_m]} Y \, d[M]\Big)$$

for every positive predictable process Y. The domination property expressed by the definition in Section 2.3 can therefore be written for every $\mathfrak{L}(\mathbf{H}; \mathbf{K})$-valued predictable process Y ($\mathbf{K}$ being an Hilbert space):

$$E\left(\left\|\int Y\,dM\right\|_{\mathbf{K}}^{2}\right) \leqslant E\left(\int_{]0,\,t_m]} \|Y\|^2\,d[M]\right).$$

From the proof of the proposition in Section 2.8, we can even write for every stopping time u,

$$E\left(\left\|\int Y\,dM^{u}\right\|^{2}\right) \leqslant 2E\left(\int_{]0,\,u]} \|Y\|^2\,d[\overline{M}]^{u}\right) + 2E\left(\|Y_u\|^2\|M_u - M_{u^-}\|^2\right).$$

Since $\|M_u - M_{u^-}\|^2 = [M]_u - [M]_{u^-}$ follows from the definition of the quadratic variation, we obtain

$$E\left(\left\|\int Y\,dM^{u}\right\|^{2}\right) \leqslant 4E\left(\int_{]0,\,u]} \|Y\|^2\,d[M]\right). \tag{4.5.1}$$

Noticing that M^u is the process $\overline{M}^u$ stopped strictly before u, we see that the same inequality holds for processes M which are *locally square integrable* martingales, i.e., processes for which there exists an increasing sequence (u_n) of stopping times such that $u_n \uparrow +\infty$ a.s. and $\overline{M}^{u_n}$ is a square integrable martingale.

Let us now remark that for every cadlag adapted Banach-valued process X, with paths of finite variation $|X|_t$ on $[0, t]$, the Schwarz inequality and the possiblity of considering $\int_{[0,\,t_m]} Y\,dX$ as a Stieltjes integral on each path (see Section 2.2) give immediately

$$E\left(\left\|\int_{[0,\,t_m]} Y\,dX^{u}\right\|^{2}\right) \leqslant E\left(|X|_u \int_{]0,\,u]} \|Y\|^2\,d|X|_u\right).$$

We may summarize this in the following proposition.

Proposition *Let X be a cadlag adapted* $\mathbf{H}$*-valued* ($\mathbf{H}$*: Hilbert) process, which may be written $X = M + N$, where M is a locally square integrable martingale and N a process with finite variation. Then if we set*

$$A := 8[M] + 2|N| + 1$$

where $|N|_t$ is the variation of N on $[0, t]$, the following inequality holds for every prelocally bounded process Y and every stopping time u:

$$E\left(\left\|\int_{[0,\,t_m]} Y\,dX^{u}\right\|^{2}\right) \leqslant E\left(A_u \int_{]0,\,u]} \|Y\|^2\,dA\right). \tag{4.5.2}$$

If $N = 0$, the same inequality holds with

$$A := 4[M] + 1.$$

A THEOREM OF GIRSANOV TYPE

In [Gir-1], it was proved that a process X of the form $M + A$, where M is a brownian motion and A some process of finite variation, becomes itself a brownian motion under a suitable change of probability. We study here (see Section 4.7 below) the effect of an absolutely continuous change of probability on processes $M + A$ where M is a square integrable martingale and A a process with finite variation.

4.6 Integration by Parts Formula

Proposition *Let* $\mathbf{H}$ *be a Hilbert space. Let* A *be an* $\mathbf{H}$*-valued cadlag adapted process with* $A_0 = 0$. *We assume that the total variation of* A *is finite on* $[0, t_m]$ *(for every sample path). Let* Z *be a real cadlag* π*-process. Then*

$$A_t Z_t = \int_{]0,t]} A_{s^-}\, dZ_s + \int_{]0,t]} Z_s\, dA_s \qquad \textit{for all} \quad t \in [0, t_m],$$

where this last integral is a Stieltjes integral (for every sample path), but may not be a stochastic integral (the process Z *is not necessarily a predictable process!)*

Proof The process A being of bounded variation, from Sections 4.3 and 4.1 we easily see that $[Z, A]$ reduces to

$$[Z, A]_t = \sum_{s \leqslant t} (\Delta Z_s)(\Delta A_s) = \int_{]0,t]} (Z_s - Z_{s^-})\, dA_s,$$

where ΔX_s denotes the jump $(X_s - X_{s^-})$ in s of the process X. Therefore, we derive from the definition of $[Z, A]$ (see 4.1.4):

$$\begin{aligned} A_t Z_t &= \int_{]0,t]} A_{s^-}\, dZ_s + \int_{]0,t]} Z_{s^-}\, dA_s + \int_{]0,t]} (Z_s - Z_{s^-})\, dA_s \\ &= \int_{]0,t]} A_{s^-}\, dZ_s + \int_{]0,t]} Z_s\, dA_s. \end{aligned}$$

4.7 Theorem

Theorem *Let* $\mathbf{H}$ *be a Hilbert space. Let* M *be an* $\mathbf{H}$*-valued cadlag adapted* π*-process and* Z *a strictly positive cadlag adapted* π*-process on* $[0, t_m]$. *We assume that* M *and* Z *are locally square integrable martingales (see Sections* 1.12 *and* 1.17*) with* $M_0 = 0$. *Then there exist an* $\mathbf{H}$*-valued adapted cadlag process of bounded variation* A *such that* $(M - A)Z$ *is locally a square integrable martingale.*

Proof From the assumption of strict positivity for Z we first derive that $1/Z$ is prelocally bounded. Let us indeed consider the increasing sequence

(v_n) of stopping times defined by $v_n := \inf\{t : Z < 1/n\} \wedge t_m$. If we set $v := \sup_n v_n$, the right continuity of Z implies that $Z_v = 0$ if $v < t_m$. Therefore $v \geqslant t_m$ and $1/Z_{v_n^-} \leqslant n$, while $\lim_n P[v_n < t_m] = 0$. The random Stieltjes integral $\int_{]0,t]} (1/Z_s)\, d[M, Z]_s$ has therefore a clear meaning and we may define an increasing cadlag process A by setting

$$A_t := \int_{]0,t]} (1/Z_s)\, d[M, Z]_s.$$

From the proposition of Section 4.6, it follows that

$$AZ = \left(\int A_{s^-}\, dZ_s\right) + \left(\int Z_s\, dA_s\right) = \left(\int A_{s^-}\, dZ_s\right) + [M, Z],$$

which implies that

$$(M - A)Z = \left(\int M_{s^-}\, dZ_s\right) + \left(\int Z_{s^-}\, dM_s\right) - \left(\int A_{s^-}\, dZ_s\right).$$

But since the processes (M_{s^-}), (Z_{s^-}), and (A_{s^-}) are locally bounded, $(M - A)Z$ is a locally square integrable martingale as a consequence of the proposition in Section 2.6.

4.8 Absolutely Continuous Change of Probability

Let $(\Omega, \mathcal{F}, P, (\mathcal{F}_t)_{t \in T})$ be a probabilized stochastic basis and Q be a probability on $(\Omega, \mathcal{F})$, equivalent to P. We assume $(\mathcal{F}_t)$ to be right continuous. We denote by Z_∞ the Radon–Nikodym derivative of Q with respect to P. If M (resp. M') is a martingale for the probability P (resp. Q), we say that M (resp. M') is a P-martingale (resp. Q-martingale).

Let X be a Hilbert-valued process which can be written $X = M + B$, where M is a locally square integrable martingale and B is a process with finite variation. We call Z the positive cadlag version (see Section 1.17) of the martingale $(E(Z_\infty \mid \mathcal{F}_t))_{t \in [0, t_m]}$. Such a cadlag version exists as a consequence of the right-continuity of $(\mathcal{F}_t)_{t \in T}$: let $\delta > 0$ and $F := \limsup_n \{Z_{t+(1/n)} - Z_t > \delta\}$. Since $F \in \mathcal{F}_t$, we have $E(1_F \cdot (Z_{t+(1/n)} - Z_t)) = 0$ for all n, which implies that $P(F) = 0$. Considering, in the same way, $F' := \limsup_n \{Z_{t+(1/n)} - Z_t > -\delta\}$, we obtain $P(F') = 0$ and therefore the right stochastic continuity of Z. We may then apply (1.17), which gives the existence of the cadlag version for Z.

Let us assume for a moment that $E|Z_\infty|^2 < \infty$. The process Z is a square integrable martingale, and we may apply the theorem of Section 4.7. There exists a process A with finite variation such that $(M - A)Z$ is a P-locally square integrable martingale. In other words, $M - A$ is Q-locally a martingale and X can be written $X = M' + A + B$, where M' is Q-locally a martingale and $A + B$ is a process with finite variation.

Remark The assumption $E|Z_\infty|^2 < \infty$ is made only to ensure that Z is a square integrable martingale. Let us agree to call a process M which is locally a P-martingale (according to the general definitions in Section 1.16 of a "local" property) a P-local martingale.

It will be seen in Section 10 that local martingales are also π-processes. Therefore it will become possible to substitute local martingales to locally square integrable martingales in the statement in Section 4.7 and to drop the assumption $E|Z_\infty|^2 < \infty$. Then we shall immediately obtain the following statement:

If P and Q are two equivalent probabilities on the stochastic right continuous basis $(\Omega, \mathcal{F}, (\mathcal{F}_t)_{t\in T})$ and if X is a Hilbert-valued cadlag process of the form $X = M + B$, where M is a P-local martingale and B is with finite variation, then $X = M' + C$, where M' is a Q-local martingale and C has finite variation.

BROWNIAN MOTION

4.9 Martingales with Bounded Variation

Proposition *Let $(M_t)_{t\in T}$ be a Hilbert-valued continuous process which is at the same time a square integrable martingale and a process with bounded variation. Then $M - M_0$ is an evanescent process.*

Proof Let s and t be two elements of T with $s < t$. Since M is a continuous process with bounded variation, $[M, M]_t - [M, M]_s = 0$ (see Section 4.3). On the other hand, since M is a martingale, we have (see Section 4.4)

$$E[\|M_t - M_s\|^2] = E\{[M,M]_t - [M,M]_s\} = 0,$$

and this proves that $M_t = M_s$ a.s. for all t.

4.10 A Characterization of the Real Brownian Motion

Theorem *Let $(W_t)_{t\in T}$ be a real continuous square integrable martingale. We assume that $T = [0, t_m]$. We denote by V the process defined by $V_t = \sigma^2 t$ for every element ω of Ω. Then, the three following properties are equivalent:*

(i) *W is a brownian motion adapted to the given stochastic basis (see Section 1.5) with $E[(W_t - W_s)^2] = \sigma^2(t - s)$ for $s < t$;*

(ii) *V is the quadratic variation of W;*

(iii) *$(W_t^2 - V_t)_{t\in T}$ is a martingale.*

Proof If W is a brownian motion, we have

$$E\{(W_t^2 - W_s^2) \mid \mathcal{F}_s\} = E[(W_t - W_s)^2 \mid \mathcal{F}_s] \quad \text{(see Section 4.4)}$$
$$= \sigma^2(t - s),$$

and (i) implies (iii). Since W is a martingale, we also have

$$E\left[(W_t - W_s)^2 \mid \mathcal{F}_s\right] = E\{([W,W]_t - [W,W]_s) \mid \mathcal{F}_s\} \qquad \text{(see Section 4.4).}$$

Then if the property (iii) is fulfilled, $[W, W] - V$ is a continuous martingale and a process with bounded variation; thus this process is evanescent (see Section 4.9) and (iii) implies (ii).

Now, we assume that V is the quadratic variation of W. Let F be an element of $\mathcal{F}_s$ with $s < t$, $s, t \in T$. We define

$$f(u) := E\left[1_F \exp^{(ia(W_{s+u} - W_s))}\right].$$

According to the Ito formula, we have

$$\exp^{[ia(W_{s+u}-W_s)]} - 1 = \int_{]s,u]} ia \exp^{[ia(W_{s+v}-W_s)]}\, dW_{s+v} - \frac{a^2}{2}\int_{]s,u]} \exp^{[ia(W_{s+v}-W_s)]}\, dV_{s+v}.$$

But, according to Section 4.2-(2), this implies that

$$\begin{aligned} f(u) - P(F) &= -\frac{a^2\sigma^2}{2} E\left\{1_F \int_{]s,u]} \exp^{(ia(W_{s+v}-W_s))}\, dv\right\} \\ &= -\frac{a^2\sigma^2}{2} \int_{]s,u]} E\left\{1_F \exp^{(ia(W_{s+v}-W_s))}\right\} dv \\ &= -\frac{a^2\sigma^2}{2} \int_{]s,u]} f(v)\, dv. \end{aligned}$$

This implies that

$$f(u) = P(F) \exp^{(-(a^2\sigma^2 u)/2)}.$$

Thus, we have

$$E\left[\exp^{(ia(W_{s+u}-W_s))} \mid \mathcal{F}_s\right] = \exp^{(-(a^2\sigma^2 u)/2)}.$$

This proves that the random variable $(W_{s+u} - W_s)$ is independent from the σ-algebra $\mathcal{F}_s$ and is gaussian centered with variance $\sigma^2 u$. Thus W is a brownian motion.

4.11 Hilbert-Valued Brownian Process

For an $\mathbf{H}$-valued process X, the variance of $X_t - X_s$, if it exists, is defined by $E[(X_t - X_s)^{\otimes 2}]$. If $E(\|X_t - X_s\|^2) < \infty$, this variance exists as an element of $\mathbf{H} \hat{\otimes}_2 \mathbf{H}$ (see Section 3.4). It is even an element of the space $\mathbf{H} \hat{\otimes}_1 \mathbf{H}$ (see Section 3.9-(2)). This comes from the equality

$$\|x \otimes y\|_{\mathbf{H}\hat{\otimes}_1\mathbf{H}} = \|x \otimes y\|_{\mathbf{H}\hat{\otimes}_2\mathbf{H}} = \|x\| \cdot \|y\|,$$

which implies that

$$E(\|X_t - X_s\|^{\otimes_2})_{\mathbf{H}\hat{\otimes}_1\mathbf{H}} = E(\|X_t - X_s\|^2) < \infty.$$

This property of the covariance is often referred to as the "nuclearity" of the covariance.

If h and g are elements of $\mathbf{H}$ and if we take the scalar product $\langle E[(X_t - X_s)^{\otimes_2}], h \otimes g\rangle$ in $\mathbf{H} \hat{\otimes}_2 \mathbf{H}$, it is immediately seen that this scalar is nothing but the covariance of the real variables $(X_t - X_s)h$ and $(X_t - X_s)g$. If $C \in \mathbf{H} \hat{\otimes}_1 \mathbf{H}$, we denote this by $C(h \otimes g)$ instead of $\langle C, h \otimes g\rangle$.

We can now give the definition of a brownian motion with values in the Hilbert space $\mathbf{H}$ and with covariance C.

Definition The $\mathbf{H}$-valued process W is called a brownian motion adapted to the given stochastic basis with covariance C if

(i) *for every* $s < t$, $W_t - W_s$ *is independent of* $\mathcal{F}_s$;

(ii) *for every* $s < t$ *and* $h \in \mathbf{H}$, *the real random variable* $(W_t - W_s)h$ *is Gaussian centered with variance* $(t - s)C(h \otimes h)$.

It is therefore immediate from the definition that the $\mathbf{H}$-valued adapted process W is a brownian motion with covariance C iff for every $h \in \mathbf{H}$, $W_t h$ is a real brownian motion with variance $C(h \otimes h)$. Therefore, we immediately obtain the following consequence of Theorem 4.10.

Proposition *The* $\mathbf{H}$*-valued continuous adapted process* W *is a brownian process adapted to the given stochastic basis, with covariance* $C \in \mathbf{H} \hat{\otimes}_1 \mathbf{H}$, *iff it is a square integrable martingale with tensor-quadratic variation*

$$[W]_t = tC.$$

EXTENSIONS AND EXERCISES

1 Let $(W_t)_{t \in T}$ be a real brownian motion with $T = [0, 1]$. For every integer n, let Y^n be the real process defined by

$$Y^n := \sum_{k=0}^{2^{-n}-1} \frac{W_{(k+1)2^{-n}} - W_{k2^{-n}}}{|W_{(k+1)2^{-n}} - W_{k2^{-n}}|} 1_{]k2^{-n}, (k+1)2^{-n}]}.$$

Is Y^n an adapted process? Is Y^n a predictable process? Is Y uniformly bounded? Prove that the sequence of random variables

$$Z^n := \int_{]0, 1]} Y^n \, dX$$

goes to infinity almost surely when n goes to infinity.

This exercise shows that it is not possible to construct the stochastic integral, as in Section 2, for processes Y which are only measurable with respect to the σ-algebra $\mathcal{F} \otimes \mathcal{T}$ (where $\mathcal{T}$ is the σ-algebra of the Borel sets of T).

2 (See [WoZ] and [All].) Let $(\Omega, \mathfrak{F}, P(\mathfrak{F}_t)_{t\in T})$ be a stochastic basis with $T := [0, 1]$. Let $(B_t)_{t\in T}$ be a real brownian motion. Let n be an integer. Let $(B_t^n)_{t\in T}$ be the process defined by $B^n := 0$, and for each element t of $[k2^{-n}, (k+1)2^{-n}]$,

$$B_t^n := B_{k2^{-n}} + 2^n(t - k2^{-n})\left[B_{(k+1)2^{-n}} - B_{k2^{-n}}\right].$$

Let σ be a real function, defined on the real line, such that its third derivative is uniformly bounded. Let X^n be the continuous process defined by

$$X_t^n := \int_0^t \sigma(B_s^n)\, dB_s^n$$

(this integral being defined, as usual, for each element ω of Ω).

The problem is to see whether the sequence of processes $(X_n)_{n>0}$ converges, when n goes to infinity, to the continuous process X defined by

$$X_t := \int_0^t \sigma(B_s)\, dB_s$$

(this integral being usual stochastic integral).

(1) For each integer n is the process B^n adapted? Is this process continuous?

(2) When n goes to infinity, does the sequence of processes $(B_n)_{n>0}$ converge uniformly for every sample function to the process B?

(3) For each integer n we define the process A^n by

$$A_t^n := \sum_{k=0}^{2^{-n}-1} B_{k2^{-n}} 1_{]k2^{-n}, (k+1)2^{-n}]}.$$

Is the process A^n adapted? Is the process A^n preditable? Is the process A^n continuous? Does the sequence of processes $(A^n)_{n>0}$ converge uniformly for every sample function to the process B when n goes to infinity?

(4) We consider the following processes:

$$C_t^n := \int_0^t \sigma(A_s^n)\, dB_s^n;$$

$$D_t^n := \int_0^t \sigma'(A_s^n)(B_s^n - A_s^n)\, dB_s^n;$$

$$R_t^n := \int_0^t \sigma(B_s^n)\, dB_s^n - C_t^n - D_t^n.$$

Study the convergence of the sequences of processes $(C^n)_{n>0}$, $(D^n)_{n>0}$ and $(R^n)_{n>0}$ when n goes to infinity.

Indication: Calculate $D^n_{(k+1)2^{-n}} - D^n_{k2^{-n}}$ and find an adequate bound for $R^n_{(k+1)2^{-n}}$. What does that mean?

3 Let M be a real square integrable martingale and A a process with finite variation on $T = [0, t_m]$.

(1) If X is the two-dimensional valued process $X := (M, A)$, what is $[X]$? Use this expression of $[X]$ to apply the Ito formula to

$$\phi_\lambda(X) := \exp\left(\lambda M - (\lambda^2/2)A\right),$$

when λ is a real number.

(2) When M is continuous and $A = [M]$, $\phi_\lambda(X)$, as defined above, is a locally square integrable martingale. Assuming moreover that

$$E\left(\int_0^{t_m} C^{2\lambda M_s}\, dA_s\right) < \infty,$$

then $\phi_\lambda(X)$ is a square integrable martingale.

(3) Conversely, if $\phi_\lambda(X)$ is a martingale for $\lambda \in [0, \lambda_0]$, $\lambda_0 > 0$, and $E[\exp^{(\lambda_0 |M_t|)}] < \infty$ for all $t \in [0, t_m]$, then M is a martingale and $[M] = A$ (to prove this last part, differentiate with respect to λ under the sign the equality

$$\int_F \exp\left(\lambda M_s - \frac{\lambda^2}{2} A_s\right) dP = \int_F \exp\left(\lambda M_t - \frac{\lambda^2}{2} A_t\right) dP,$$

for $F \in \mathcal{F}_s$, $s < t$, and derive from there

$$\int_F M_s\, dP = \int_F M_t\, dP$$

and

$$\int_F (M_s^2 - A_s)\, dP = \int_F (M_t^2 - A_t)\, dP\Big).$$

HISTORICAL NOTES

Section 3. The Ito formula is, properly speaking, the formula of change of variable for processes that are integrals of the brownian motion ([Ito-3]). The successive extensions to more general situations were made by Kunita and Watanabe, Meyer, and Doléans-Dade (see [KuW], [DoM-1], [Mey-3], [Mey-4]). The introduction of the tensor quadratic variation by the authors (see [Met-5]) allowed them to consider Hilbert-valued and even Banach-valued processes (see [GrP]). The proof given here differs rather deeply from known ones inasmuch as it does not use any decomposition of the processes into their martingale part and some "increasing part," but relies directly on the π-domination property and on the related existence of stochastic integrals.

Section 4. Since the Ito formula is the corner stone of stochastic "differential calculus," its applications are countless. We have chosen only a few of them here, in particular to illustrate the use of the tensor mutual variation of two processes. The form given here of the "Girsanov theorem" is due to Van Schuppen and Wong [VsW] for local martingales and Jacod and Memin [JaM] for semimartingales.

CHAPTER 3

STOCHASTIC INTEGRAL EQUATIONS

The concept of stochastic differential (or integral) equations has been present in the literature for a long time, even before the first systematic treatment by K. Ito, to whom are due most of the basic ideas of the theory (cf. [Ito]).

In this chapter, we first give classical examples (Section 5) and then present a theory for general stochastic integral equations (Section 6) with a Lipschitz-type hypothesis. In these equations, the stochastic driving term (which plays the role of brownian movement in Ito's equations) belongs to a class of processes—we call them π^*-processes—that satisfy an inequality apparently stronger than the one defining the π-processes. However, it should be mentioned at once that the class of π^*-processes will later turn out to coincide with the class of π-processes (at least for real processes) and contain all the processes which have been considered up to now for the purpose of stochastic integration and stochastic equations (cf. [Mey-4], [DoM-2], [Pro-1, 2], [Eme]). The proof of this important fact is given in Section 10.

In this chapter, we do not want to bother the reader by studying to what extent the class of processes considered here is general. For the reader's ease, we shall only prove that this class contains sufficiently many "basic" processes (such as brownian motion, Poisson processes, all processes with finite variation, and continuous martingales) and is stable through operations such as localization and change of variables. This already makes it big enough for most users. We prove a basic existence and uniqueness theorem (Section 6.8) for this class of driving stochastic terms.

Finally, Section 7 is devoted to establishing a condition of nonexplosion for solutions and to studying the stability of the solutions under some "small" perturbation of the coefficients, initial conditions, and driving term.

The notion of smallness of the perturbation of the driving term needs the consideration of some topology on the space of π^*-processes: this topology is introduced in Section 7.

5 EXAMPLES OF STOCHASTIC DIFFERENTIAL EQUATIONS

5.1 Langevin Equation and Orstein–Uhlenbeck Process

We mention an old example, connected with the Ornstein–Uhlenbeck theory for a free particle as quoted in [Nel]. The stochastic equation, which Nelson refers to as the Langevin equation, is

$$\frac{d^2X}{dt^2} + \alpha \frac{dX}{dt} = \sigma^2 \frac{d\beta}{dt}, \tag{5.1.1}$$

which governs the movement of a particle with coordinate X, submitted to infinitely many uncorrelated small shocks represented by "the increments $d\beta_t$ at time t" of the brownian motion β.

Written in this way, the above equation has no immediately clear meaning because of the nondifferentiability of β. However, if we consider the velocity $dX/dt := V$, we can say that X is a process solution of the above equation if there exists a process V such that

(i) $$X(t) = X_0 + \int_0^t V(s)\,ds \qquad \text{(pathwise)},$$

(ii) $$V(t) - V(0) + \alpha \int_0^t V(s)\,ds = \sigma^2\beta(t) \qquad \text{(equality of processes up to } P\text{-equivalence)},$$

where $V(0)$ and X_0 are initial values; this is perfectly meaningful when assuming that a stochastic basis and a brownian motion β on this basis are given.

It is clear from the above definition that if V and V' are two solutions of the equation (ii), they satisfy

$$V(t) - V'(t) + \alpha \int_0^t (V(s) - V'(s))\,ds = 0 \qquad \text{on } P\text{-almost all paths},$$

and therefore are P-equivalent.

This expresses the uniqueness, up to P-equivalence, of the solutions of (i) and (ii).

Setting

$$\phi(x, y) = xy,$$

and applying the Ito formula of Section 3.3 in the form

$$\phi(M_t, U_t) = \phi(M_0, U_0) + \int_0^t \frac{\partial\phi}{\partial x}(M_s, U_s)\, dM_s$$

$$+ \int_0^t \frac{\partial\phi}{\partial y}(M_s, U_s)\, dU_s + \frac{1}{2}\int_0^t \frac{\partial^2\phi}{\partial x\,\partial y}\, d[M_s, U_s],$$

with the processes

$$M_t := \int_0^t \sigma e^{-\alpha s}\, d\beta_s,$$

$$U_t := e^{\alpha t},$$

we obtain (note that $[M, U] = 0$ according to Proposition 4.3)

$$M_t + U_t = \sigma\beta_t + \int_0^t \alpha\left(\int_0^s \sigma e^{\alpha(s-u)}\, d\beta_u\right) ds.$$

If we set

$$V_t := e^{\alpha t}V_0 + \int_0^t \sigma e^{\alpha(t-u)}\, d\beta_u = e^{\alpha t}V_0 + M_t U_t,$$

we get from the above formula

$$V_t = e^{\alpha t}V_0 + \sigma\beta_t + \alpha\int_0^t V_s\, ds - \alpha\int_0^t e^{\alpha s}V_0\, ds,$$

or

$$V_t = \sigma\beta_t + \alpha\int_0^t V_s\, ds + V_0,$$

which shows V to be solution of (ii).

The thus defined process V is called the *Ornstein–Uhlenbeck velocity process*, while the corresponding position process X is called the *Ornstein–Uhlenbeck process*.

We leave it to the reader to prove that V is a Gaussian process (i.e., all finite-dimensional distributions $(V_{t_1} \cdots V_{t_n})$ are Gaussian) and that the covariance function $C(s, t)$ is given for $s \leqslant t$ by the formula

$$C(s,t) := E(V(s)V(t)) = \int_0^s e^{\alpha(t-u)}\sigma^2 e^{\alpha(s-u)}\, du.$$

5.2 Ito equations

In (5.1.1) the "random perturbation" $\sigma^2\, d\beta_t$ depends only on the time and not on the position X of the particle. In the more general situation studied by Ito, the "perturbation" is of the form $\sigma_t^2(X)\, d\beta_t$, where $\sigma_t(\cdot)$ is a function of t and x. More precisely, the Ito equations are of the form:

$$dX(t) = \alpha(t, X(t))\, dt + \sigma^2(t, X(t))\, d\beta_t, \tag{5.2.1}$$

where for every t, $\alpha(t, \cdot)$ is a mapping of $\mathbf{R}^n$ into $\mathbf{R}^n$, $\sigma^2(t, \cdot)$ a mapping of $\mathbf{R}^n$ into $\mathfrak{L}(\mathbf{R}^m; \mathbf{R}^n)$, and β is an $\mathbf{R}^m$-dimensional brownian motion given on a stochastic basis.

A solution of (5.2.1) (or to put it better, a "strong solution" as opposed to "weak solutions," as in the case considered in [StV]; see also [Pri]) with initial condition X_0 is a process X on the given stochastic basis such that

$$X(t) = X_0 + \int_0^t \alpha(s, X(s))\, ds + \int_0^t \sigma^2(s, X(s))\, d\beta_s. \tag{5.2.2}$$

If (5.2.2) defines X up to P-equivalence, the solution is unique.

The study of the existence and uniqueness of the solution of (5.2.2) will appear as a particular case of the study which will be made in the next section.

5.3 Doléans-Dade–Protter equations

Equations have been considered by Doléans-Dade and Protter separately [Dol-4], [Pro-1, 2], which contain a stochastic driving term that is not brownian but can be a general "semimartingale," i.e., a π-process in the finite-dimensional case (see Section 2).

If Z is an n-dimensional π-process, such an equation can be written

$$dX(t) = f(t, X\ \ (t))\, dZ(t), \qquad f(t, x) \in \mathfrak{L}(\mathbf{R}^m; \mathbf{R}^n), \tag{5.3.1}$$

where X_- denotes the cadlag process obtained from X by setting

$$X_-(\omega, t) = \lim_{s \uparrow t,\, s < t} X_s(\omega, s).$$

A solution of this equation with initial condition X_0 is a cadlag n-dimensional adapted process X such that

$$X(t) = X_0 + \int_0^t f(s, X_-(s))\, dZ(s), \tag{5.3.2}$$

where X_0 is an n-dimensional $\mathfrak{F}_0$-measurable random variable.

One may also consider equations of the form

$$X(t) = V(t) + \int_0^t f(s, X_-(s))\, dZ(s),$$

where V is a given adapted cadlag process as in [Mey-5].

It should be noted that since Z is a general π-process, the integral in (5.3.2) is meaningless unless $(f(s, X_-(s)))_{s\in T}$ is a predictable process. This is the case when $x \rightsquigarrow f(s, x)$ is continuous, the process $(f(x, X_-(s)))_{s\in T}$ being then left continuous. General conditions will be given in Section 6 for the predictability of $f(\cdot, X)$.

Equation (5.3.2) will appear as a special case of those considered in the next Section.

6 GENERAL STOCHASTIC INTEGRAL EQUATIONS

DESCRIPTION OF THE EQUATIONS, STRONG SOLUTIONS

6.1 Generalities

The equations to be considered in this section will be written

$$dX(t) = dV(t) + a(X,t)\,dZ(t)$$

or in the integral form

$$X(t) = V(t) + \int_{]0,\,t]} a(X,s)\,dZ(s), \tag{6.1.1}$$

where V is a cadlag process. The process Z and the functional a of the process X will be described now. In particular, the hypothesis on Z and a will give a meaning to the stochastic integral in (6.1.1), and the notion of the solution of (6.1.1) will be given a precise definition.

We fix a few notations and assumptions which hold throughout this section:

T is a bounded closed interval $[0, t_m]$ of the real line;

$\mathbf{H}, \mathbf{K}$ are two separable Banach spaces and $\mathcal{L}(\mathbf{K}, \mathbf{H})$ is the Banach space of continuous linear operators with the usual norm of bounded operators; in the sequel the norms on $\mathbf{H}$, $\mathbf{K}$, and $\mathcal{L}(\mathbf{K}, \mathbf{H})$ will be denoted by $\|\cdot\|$; let us recall that for all $k \in \mathbf{K}$ and $u \in \mathcal{L}(\mathbf{K}, \mathbf{H})$ we have

$$\|u(k)\| \leqslant \|u\|\,\|k\|;$$

$(\Omega, \mathcal{F}, P, (\mathcal{F}_t)_{t\in T}) = \mathbf{B}^I$ is a complete stochastic basis (see Section 1.1), the family $(\mathcal{F}_t)_{t\in T}$ being assumed right continuous. We shall call this basis the "initial basis." We denote by $\mathcal{Q}$ the algebra generated by the sets $F \times]s, t]$ with $F \in \mathcal{F}_s$; the σ-algebra generated by $\mathcal{Q}$ is the σ-algebra of predictable sets (see Section 1).

6.2 Predictable Functions of a Process—Canonical Basis

The functional a occurring in (6.1.1) will be defined in such a way that for a cadlag adapted process X, $(a(X,s) : s \in T)$ is a predictable process.

In order to define the predictable functionals of a process, we introduce the following spaces.

Let $\mathbf{H}$ be a Banach space; we will denote by $\mathbf{D}^{\mathbf{H}}$ the set of all the mappings from T into $\mathbf{H}$ which are cadlag (i.e., those that according to the definitions of Section 1.4 are right continuous, have left limits in every $t \in T$).

For every $t \in T$, $\mathcal{D}_t^{\mathbf{H}}$ denotes the σ-algebra generated by the "cylinders" $\{f : f \in \mathbf{D}^{\mathbf{H}}, f(s) \in B\}$, where $s \leqslant t$ and B is any Borel set in $\mathbf{H}$.

We set $\tilde{\Omega}^{\mathbf{H}} := \mathbf{D}^{\mathbf{H}} \times \Omega$, and we consider on $\tilde{\Omega}^{\mathbf{H}}$ the increasing family $(\tilde{\mathcal{F}}_t^{\mathbf{H}})_{t \in T} := (\mathcal{D}_t^{\mathbf{H}} \otimes \mathcal{F}_t)_{t \in T}$ of σ-algebras.

$(\tilde{\Omega}^{\mathbf{H}}, \tilde{\mathcal{F}}_t^{\mathbf{H}})$ is a stochastic basis which will be denoted by $\mathbf{B}^{\mathbf{H}}$ and called the canonical basis (for the $\mathbf{H}$-valued processes).

According to the general definitions of Section 1, the σ-algebra of subsets of $\tilde{\Omega}^{\mathbf{H}}$ generated by the family $\{\tilde{F} \times]s, t] : s \leqslant t \in T, \tilde{F} \in \tilde{\mathcal{F}}_s^{\mathbf{H}}\}$ is called the σ-algebra of predictable subsets of $\tilde{\Omega}^{\mathbf{H}}$. It will be denoted by $\tilde{\mathcal{P}}^{\mathbf{H}}$. A mapping a: $\tilde{\Omega}^{\mathbf{H}} \to \mathbf{L}$, where $\mathbf{L}$ is a Banach space, will be said to be an $\mathbf{L}$-valued predictable functional on the $\mathbf{H}$-valued cadlag adapted processes if it is a measurable mapping from $(\tilde{\Omega}^{\mathbf{H}}, \tilde{\mathcal{P}}^{\mathbf{H}})$ into $\mathbf{L}$.

6.3 Remarks and Conventions

(a) The σ-algebra of predictable sets of the canonical basis $\mathbf{B}^{\mathbf{H}}$ is generated by the sets $G \times F \times]s, t]$, where G is an element of $\mathcal{D}_s^{\mathbf{H}}$ and F is an element of $\mathcal{F}_s$; actually, it is sufficient to consider the sets G of the following form: $G = \{x : x \text{ cadlag}, x(u) \in \mathbf{H}_0\}$, with $u < s$ and $\mathbf{H}_0$ a Borel set of $\mathbf{H}$.

(b) Let $a(x, \omega, t)$ be an $\mathcal{L}(\mathbf{K}, \mathbf{H})$-valued process defined with respect to the canonical basis $\mathbf{B}^{\mathbf{H}}$. Let X be an $\mathbf{H}$-valued process defined with respect to the initial basis $\mathbf{B}^I$. In the sequel we consider processes Y such that $Y_t(\omega) = a[X(\omega), \omega, t]$; in this situation, in accordance with traditional notations, we dispense with writing the symbol ω when we consider the random variable Y_t, and we write $Y_t = a(X, t)$ or $Y_t = a(X, \cdot, t)$.

(c) Since X is a cadlag process, we shall make a systematic use of the notation ΔX for the process defined by

$$\Delta X_t(\omega) = X_t(\omega) - \lim_{\substack{s \uparrow t \\ s < t}} X_s(\omega).$$

Thus for every stopping time u, ΔX_u is the jump of X at time u.

(d) Let u be a stopping time and X and Y be two processes; the process X is cadlag. Then, we write

$$\sup_{t < u} \left\| \int_{]0, t]} Y_s \, dX_s \right\|^2$$

for the random variable U defined more precisely by

$$U(\omega) = \sup_{t < u(\omega)} \|Z_t(\omega)\|^2,$$

where Z is the cadlag integral process of Y with respect to X (defined up to

P-equivalence (see Section 2) and denoted by $(\int Y\,dX)$). We recall the notation

$$Z_t = \int_{]0,\,t]} Y_s\,dX_s,$$

or

$$Z_t = \int_0^t Y_s\,dX_s$$

if there is no possible ambiguity, in particular if X is continuous (and therefore Z: see Section 2.5).

The following proposition shows that whenever a is a predictable functional and X a cadlag adapted process, the process $a(X, \cdot t)$ occurring in the stochastic integral in (6.1.1) is predictable, which gives meaning to this integral when Z is any π-process.

6.4 A Proposition

Proposition *Let* $\mathbf{L}$ *be a Banach space. Let* X *be a cadlag* $\mathbf{H}$*-valued process defined and adapted with respect to the initial basis* $\mathbf{B}^I$. *Let* $a(x, \omega, t)$ *be an* $\mathbf{L}$*-valued process defined and predictable with respect to the canonical basis* $\mathbf{B}^\mathbf{H}$. *Let* Y *be the process defined by* $Y_t(\omega) := a(X(\omega), \omega, t)$. *Then* Y *is an* $\mathbf{L}$*-valued process, predictable with respect to the initial basis* $\mathbf{B}^I$. *Moreover* $Y_t(\omega)$ *depends only on the values* $X_s(\omega)$ *for* $s < t$ *(it is therefore possible to define* $Y_t(\omega)$ *when* X_s *is known only for* $s < t$).

Proof (1) First we consider the case for which there exists a k in $\mathbf{L}$, $u < v < w$ are elements of T, $\mathbf{H}_0$ is a Borel subset of $\mathbf{H}$, and F is an element of $\mathcal{F}_v$ such that if

$$J := \{x : x_u \in \mathbf{H}_0\},$$

then

$$a(x, \omega, t) = k 1_J(x) 1_F(\omega) 1_{]v,\,w]}(t).$$

Let F' be the set defined by

$$F' := \{\omega : X_u(\omega) \in \mathbf{H}_0\}.$$

Since the process X is adapted, F' belongs to $\mathcal{F}_u$; we also have

$$\begin{aligned} Y_t(\omega) = a(X(\omega), \omega, t) &= k 1_J(X(\omega)) 1_F(\omega) 1_{]v,\,w]}(t) \\ &= k 1_F(\omega) 1_{F'}(\omega) 1_{]v,\,w]}(t). \end{aligned}$$

Then Y is a predictable process and $Y_t(\omega)$ depends only on $X_s(\omega)$ for $s < t$.

(2) Next we consider an $\mathbf{H}$-valued process X adapted with respect to the initial basis $\mathbf{B}^I$. Let $\mathcal{C}_X$ be the family of all the $\mathbf{L}$-valued processes a defined with respect to the canonical basis $\mathbf{B}^\mathbf{H}$ and such that if $Y = a(X, t)$, Y is a predictable process with Y_t depending only on X_s for $s < t$. The

space $\mathcal{C}_X$ is a vector space and a monotone class; moreover, $\mathcal{C}_X$ contains all the processes $a = k 1_J 1_F 1_{]v,w]}$ as defined in (1). Then $\mathcal{C}_X$ contains all the predictable processes (cf. Remark 6.3(a)).

6.5 Strong Solutions

An **H**-valued process X defined up to P-equivalence on the open (resp. closed) stochastic interval $[0,u[$ (resp. $[0,u]$) is said to be a *strong solution* of (6.1.1) on $[0,u[$ (resp. $[0,u]$) if the process $(\int_{]0,t]} a(X,s)\,dZ_s)_{t\in T}$ is well defined on $[0,u[$ (resp. $[0,u]$) as a cadlag adapted process and differs from $(X - V)$ by a P-null process.

Let us remark that according to the proposition in Section 6.4, X need only be given cadlag and adapted on $[0,u[$ with left limits at u for $a(X,s)$ in order to exist on $[0,u]$. Thus a strong solution X on $[0,u[$ with left limits at u can be extended into a strong solution X' on $[0,u]$.

This last solution is then determined by

$$X'_u = \lim_{\substack{t\uparrow u\\ t<u}} X_t + a(X,u)\,\Delta Z_u + \Delta V_u.$$

6.6 Lipschitz and Boundedness Hypothesis on a

The main hypotheses which will be made on a, besides predictability, are of Lipschitz type.

For future reference the functional a will be said to satisfy

(L_1) if there exists a cadlag adapted increasing process $L \geqslant 0$:

$$\|a(f,\omega,t) - a(f',\omega,t)\| \leqslant L_t(\omega) \sup_{0\leqslant s<t} \|f(s) - f'(s)\|,$$

$$\forall t \in T, f, f' \in D^{\mathbf{H}}$$

(uniform Lipschitz property for every sample path);

(L_2) if for every $b > 0$ there exists a cadlag adapted increasing process $L_b > 0$:

$$\|a(f,\omega,t) - a(f',\omega,t)\| \leqslant L_b(\omega,t) \sup_{0\leqslant s<t} \|f(s) - f'(s)\|,$$

$$\forall t \in T, \forall f, f' \in \left\{ g : g \in D^{\mathbf{H}},\ \sup_{0\leqslant s<t} \|g(s)\| \leqslant b \right\}$$

(local Lipschitz property);

(L_3) if the process $(a(0,\cdot,t))_{t\in T}$ is locally bounded.

It should be noted that the condition (L_3) is automatically satisfied when the process $(a(0,\cdot,t))_{t\in T}$ is cadlag as is the case in [EMe]. Moreover, if

(L_2) and (L_3) hold simultaneously, the inequality

$$\|a(f, \omega, t)\| \leqslant \|a(0, \omega, t)\| + bL_b(\omega, t) \qquad \text{where} \quad b := \sup_{0 \leqslant s < t} \|f(s)\|$$

shows that *for every cadlag adapted process X the adapted process* $(a(X, \cdot, t))_{t \in T}$ *is itself locally bounded.*

π^*-PROCESSES

6.7 Definition of π^*-Processes

Let us recall that we can define the stochastic integral process $(\int Y\, dZ)$ for all locally bounded predictable processes Y if Z is a π-process; moreover, if Z is such a π-process, we have the Ito formula.

Now, to study stochastic differential equations we use a domination property for Z which is stronger than the domination property expressed in the definition of a π-process. As a consequence of an important theorem (cf. Section 10) these two properties turn out to be equivalent in the finite-dimensional case.

Referring to a frequently used notation $X_t^* = \sup_{s \leqslant t} \|X_s\|$ (which will be used in the sequel), we introduce the notation of a *-dominating increasing process.

Definition A cadlag adapted process Z with values in a Banach space **K** will be called a π^*-process if there exists a cadlag adapted increasing process Q with the following property:

[π^*] for every Hilbert space **H**, every $\mathcal{Q}$-simple $\mathcal{L}(\mathbf{K}, \mathbf{H})$-valued process Y, and every stopping time u the following inequality holds:

$$E\left\{ \sup_{0 \leqslant s \leqslant u} \left\| \int_{]0, t]} Y_s\, dZ_s \right\|^2 \right\} \leqslant E\left\{ Q_{u^-} \int_{[0, u[} \|Y_s\|^2\, dQ_s \right\}.$$

In such a situation we say that Q *-dominates Z.

Remark 1 It is trivial from the definition that every π^*-process is a π-process.

Remark 2 The uniform approximation property of Section 2.5 shows that the inequality [π^*] holds for every locally bounded predictable process Y when Z is a π^*-process dominated by Q.

Remark 3 Noticing once more that $(\int Y\, dZ) = (\int YU\, dX)$ when $Z = (\int U\, dX)$, we readily get that the integral process $(\int U\, dX)$ is a π^*-process as soon as X is a π^*-process and U a locally bounded predictable process. If Q *-dominates X, then $(Q_t + \int_{]0, t]} \|U_s\|^2\, dQ_s)_{t \in T}$ *-dominates Z.

6.8 A Lemma on Integration

For easy future reference we state separately here a classical lemma which will be used again later and is needed now to give examples of π^*-processes.

Lemma *Let U and ϕ be positive random variables such that for every $\xi > 0$ and some $\alpha > 0$ and $p > 1$ with $p - \alpha > 0$ we have*

$$\xi^\alpha P[U > \xi] \leqslant \int_{[U>\xi]} \phi^\alpha \, dP$$

and

$$\int \phi^p \, dP < \infty \qquad \text{for some} \quad p > 1.$$

Then

$$E(U^p) \leqslant (p/(p-\alpha))^{p/\alpha} E(\phi^p).$$

Proof We recall the proof of this lemma, which consists of several successive applications of the Fubini's formula. Denoting the law of U by μ, we write

$$\begin{aligned} E(U^p) &= \int_0^\infty y^p \mu(dy) = \int_0^\infty \left(\int_0^y p\xi^{p-1} \, d\xi \right) \mu(dy) \\ &= \int_0^\infty p\xi^{p-1} \, d\xi \left(\int_{]\xi, \infty]} \mu(dy) \right) \\ &\leqslant \int_0^\infty p\xi^{p-1-\alpha} \, d\xi \left(\int_{[U>\xi]} \phi^\alpha \, dP \right). \end{aligned}$$

If now

$$F(y) := \int_{[U \leqslant y]} \phi^\alpha \, dP,$$

we may write

$$\begin{aligned} E(U^p) &\leqslant \int_0^\infty p\xi^{p-1-\alpha} \, d\xi \left(\int_{]\xi, \infty]} dF(y) \right) = p \int_0^\infty dF(y) \left(\int_0^y \xi^{p-1-\alpha} \, d\xi \right) \\ &= \frac{p}{p-\alpha} \int_0^\infty y^{p-\alpha} \, dF(y) \leqslant \frac{p}{p-\alpha} \int_0^\infty U^{p-\alpha} \phi^\alpha \, dP. \end{aligned}$$

Applying the Hölder inequality, we obtain

$$E(U^p) \leqslant (p/(p-\alpha))[E(U^p)]^{(p-\alpha)/p} E(\phi^p)^{\alpha/p},$$

and therefore the inequality of the lemma.

6.9 Examples of π^*-Processes

We can immediately see that every process Z with finite variation is a π^*-process with the variation process $|X|$ as a *-dominating process.

We cannot prove now (this is deferred to Section 10) that every martingale is a π^*-process. We restrict ourselves for the present to proving that fact for continuous martingales. This is part of the following statement which reveals something of the bigness of the class of π^*-processes.

Proposition *Let Π^* be the class of π^*-processes and $\Pi^*_{\mathbf{K}}$ the vector space of* **K**-*valued π^*-processes.*

(1) *Π^* contains all cadlag adapted Banach-valued processes with finite variation and Hilbert-valued continuous square integrable martingales.*

(2) *If $X \in \Pi^*$, the same is true for the process $\phi \circ X$, where ϕ is any twice continuously differentiable function, as in the Ito formula of Section* 3.7.

(3) *If $X \in \Pi^*_{\mathbf{K}}$, the same is true for the stopped processes $\bar{X}^u := (X_{u \wedge t})_{t \in T}$, where u is any stopping time, and for the process X^u, stopped strictly before u as well.*

(4) *Conversely, $\Pi^*_{\mathbf{K}}$ has the following "local property": if there exists an increasing sequence (v_n) of stopping times such that $\lim_n P[v_n < t_m] = 0$ and $\bar{X}^{v_n}$ or X^{v_n} belong to $\Pi^*_{\mathbf{K}}$ for all n, then $X \in \Pi^*_{\mathbf{K}}$.*

Proof We have already mentioned that the first part of statement (1) is easily checked, so we leave the proof to the reader. Statement (2) follows immediately from the application of the Ito formula and Remark 3 following the definition of π^*-processes. The same remark implies readily that the stopped process $(X_{u \wedge t})_{t \in T}$, which is the integral process $(\int 1_{]0, u]}\, dX)$, is itself a π^*-process. The same property follows for X^u as a consequence of $X^u = (\int 1_{]0, u]}\, dX) - \Delta X_u 1_{[u, \infty[}$, the process $\Delta x_u 1_{[u, \infty[}$ being a cadlag process with finite variation and therefore a π^*-process. Statement (4) follows from the Fatou lemma and the consideration of the process $Q := \sum_k 2^k Q^k 1_{]v_k, t_m]}$, where Q^k dominates X^{v_k}. Then we have only to prove the part of statement (1) which concerns continuous martingales. It is sufficient to prove the following point: there exists a cadlag adapted increasing process Q such that for every finite set $J \subset T$, and $\mathcal{C}$-simple process Y, and every stopping time u we have

$$E\left\{ \sup_{\substack{0 \leqslant t < u \\ t \in J}} \left\| \int_{]0, t]} Y_s \, dZ_s \right\|^2 \right\} \leqslant E\left\{ Q_u \text{-} \int_{]0, u[} \| Y_s \|^2 \, dQ_s \right\}. \tag{6.9.1}$$

The inequality $[\pi^*]$ will follow from (6.9.1) by considering an increasing sequence (I_n) of finite sets I_n, the union of which is dense in T.

For every $\xi > 0$ we define the stopping time

$$\tau := \inf\left\{ t : t \in T, \left\| \int_{]0,t]} Y_s \, dZ_s \right\|^2 > \xi \right\} \wedge u.$$

Let us denote the process $(\int Y \, dZ)$ by U and write

$$U^* := \sup_{\substack{0 \leqslant t < u \\ t \in J}} \| U_t \|.$$

We have

$$\{ U^* > \xi \} = \{ \tau < u \},$$

and from the definition of τ, it follows that

$$\xi P[U^* > \xi] \leqslant \int_{\{\tau < u\}} \| U_\tau \| \, dP.$$

But we know from the proposition in Section 2.6 that the process U is a continuous martingale if Z is, and the same holds for the stopped process $(U_{u \wedge t})_{t \in T}$. From Section 1.18, we also know that $\| U_{u \wedge t} \|_{t \in T}$ is a submartingale. Noticing that the stochastic interval $]\tau, t_m]$ is an element of $\mathcal{Q}$, since τ is simple, we obtain

$$\int_{\{\tau < u\}} \| U_u \| - \| U_\tau \| \, dP = E(\| U_u \| - \| U_{\tau \wedge u} \|) > 0$$

from the definition of a submartingale (see Section 1.17). Therefore, we may write

$$\xi P[U^* > \xi] \leqslant \int_{\{U^* > \xi\}} \| U_u \| \, dP,$$

and the lemma of Section 6.8 gives

$$E \| U^* \|^2 \leqslant 4 E \| U_u \|^2.$$

Introducing the process A as in the proposition in Section 4.5, and noticing that it is continuous, we see that (6.9.1) holds if we take $Q = 2A$, and this completes the proof.

EXISTENCE AND UNIQUENESS OF SOLUTIONS

6.10 Main Theorem on Existence and Uniqueness

Theorem *Let* **H** *be a Hilbert space and* **K** *a separable Banach space. Let* $\mathbf{B}^I = (\Omega, \mathfrak{F}, P, (\mathfrak{F}_t)_{t \in T})$ *be a stochastic basis with the usual assumptions (cf.* Section 6.1) *called the initial basis, and* Z *be a* **K**-*valued cadlag process defined and adapted with respect to the initial basis* $\mathbf{B}^I$. *We suppose that* Z *is* *a π^*-process with dominating increasing process Q. Let $a(x, \omega, t)$ be an $\mathfrak{L}(\mathbf{K}; \mathbf{H})$-valued process defined and predictable with respect to the canonical basis $\mathbf{B}^{\mathbf{H}}$ (see Section 6.2). We assume the local Lipschitz property* (L_2) *for a and the property* (L_3). *Let V be an $\mathbf{H}$-valued cadlag adapted process.*

Then there exists a unique stopping time v and an $\mathbf{H}$-valued cadlag process X, defined on $[0, v[$, adapted with respect to the initial basis $\mathbf{B}^I$, and unique up to P-equivalence, with the following two properties:

(i) *If ω belongs to the set $\{v < t_m\}$, $\limsup_{t \uparrow v(\omega)} \|X_t(\omega)\| = +\infty$.*
(ii) *$X_t = V_t + \int_{]0, t]} a(X, s)\, dZ_s$ on the stochastic interval $[0, v[$.*

The stopping time v is actually predictable.

Proof The proof has three steps. In Section 6.11, we prove uniqueness. In Section 6.12 the existence of the solution X is established under stronger hypotheses, namely, conditions (L_1) and (L_2) for a. The last step is performed in Section 6.14, where we shall see that in the general setting (i) expresses the "maximality" of the solution (X, v); the existence of such a "maximal solution" is there proved by using the "extension" lemma of Section 6.13.

6.11 Uniqueness

Proposition *We consider the hypothesis and notations given in Theorem* 6.10. *Let X and X' be two cadlag adapted processes for which* (ii) *holds on the stochastic intervals $]0, v]$ and $]0, v']$, respectively. Then $X 1_{[0, v \wedge v']}$ and $X' 1_{[0, v \wedge v']}$ are two P-equivalent processes.*

Proof We define

$$u := v \wedge v' \wedge \inf\{t : \|X_t - X_t'\| > 0\}.$$

If $P([u < (v \wedge v')]) = 0$, there is nothing to prove. Then let us suppose that $P([u < (v \wedge v')]) > 0$. Since the processes X and X' are cadlag, there exists a positive number b and a stopping time w' such that

$$\sup_{u \leqslant s < w'} (\|X_s\| + \|X_s'\|) \leqslant b,$$

$$P([w' > u]) > 0 \qquad \text{and} \qquad w' \leqslant (v \wedge v').$$

Let L_b be the process satisfying condition (L_2). We define a stopping time w by setting

$$w := w' \wedge \inf\{t : t \geqslant u, Q_t(Q_t - Q_u) > 1/(2L_b)\}.$$

The processes Q and L_b being right continuous, we have $P([w > u]) > 0$.

Then we define

$$h := E\left\{ \sup_{u \leqslant s < w} \|X_s - X'_s\|^2 \right\}$$

and we have

$$h = E\left\{ \sup_{u \leqslant t < w} \left\| \int_{]u,\,t]} [a(X,s) - a(X',s)]\, dZ_s \right\|^2 \right\}$$

$$\leqslant E\left\{ Q_{w-} \int_{]u,\,w[} \|a(X,t) - a(X',t)\|^2\, dQ_t \right\}$$

$$\leqslant E\left\{ Q_{w-} \int_{]u,\,w[} L_b(t) \sup_{u \leqslant t < w} \|X_t - X'_t\|^2\, dQ_t \right\} \leqslant \tfrac{1}{2} h,$$

and therefore $h = 0$, which contradicts the definition of u; then $P([u < (v \wedge v')]) = 0$, which proves the uniqueness.

6.12 Extension Lemma (Uniform Lipschitz Case)

Lemma *We consider the hypothesis and notations given in the theorem of Section* 6.10. *Let u be a stopping time and X a strong solution* (*cf. Section* 6.5) *of* (6.1.1) *on* $[0, u]$. *Moreover, we assume that the "random functional" a satisfies* (L_1) *on* $\{X_u^* \leqslant b\}$ *and* (L_3). *Let v be the stopping time defined as follows*:

if $X_u^* > b$,

$$v := u;$$

if $X_u^* \leqslant b$,

$$v := \inf\left\{ t : t \geqslant u,\ V_t^* + Q_t \int_{]0,\,t]} \|a(0,s)\|^2\, dQ_s > b \right.$$

$$\left. \text{or} \quad Q_t \int_{]u,\,t]} L_s\, dQ_s > \tfrac{1}{2} \right\} \wedge t_m,$$

where b is a positive number.

Then there exists a strong solution X' of (6.1.1) *on* $[0, v]$

Proof For every **H**-valued cadlag process W we put

$$|||W||| := \left[E\left\{ \sup_{u < t < v} \|W_t\|^2 \right\} \right]^{1/2},$$

where the supremum is taken equal to zero if $u = v$. We denote by $\mathfrak{W}$ the space of all **H**-valued cadlag adapted processes W defined up to P-equivalence on $[0, v]$ such that $|||W||| < +\infty$ and $W 1_{[0,\,u]} = X 1_{[0,\,u]}$. For every $W \in \mathcal{W}$ we define the process UW on $[0, v]$ as follows:

$$(UW)_t := V_t + \int_0^t a(W,s)\,dZ_s.$$

$$(UW)_t = (X1_{[0,u]})_t + (V1_{]u,v]})_t + \int_{]u,t]} a(W,s)\,dZ_s + X_u(1_{]u,v]})_t.$$

According to the remark following the definition of (L_3) in Section 6.6, the stochastic integral in this last formula exists since $a(W,s)$ is locally bounded. Moreover,

$$|||UW|||^2 \leqslant 2E\left\{\sup_{u<t<v}\left\|V_t + \int_{]0,t]}(a(W,s)-a(0,s))\,dZ_s\right\|^2\right\}$$

$$+ 2E\left\{\sup_{u<t<v}\left\|\int_{]0,t]} a(0,s)\,dZ_s\right\|^2\right\}$$

$$\leqslant 4E\left\{Q_{v^-}\int_{]u,v[} 3L_s\left(X_u^{*2}1_{\{u<v\}} + V_u^{*2} + \sup_{u<t<s}\|W_t\|^2\right)dQ_s\right\}$$

$$+ 2E\left\{Q_{v^-}\int_{]u,v[}\|a(0,s)\|^2\,dQ_s\right\}$$

$$\leqslant 2E\left(3b^2 + \sup_{u<t<s}\|W_t\|^2\right) + 2b \leqslant 6b^2 + 2|||W|||^2 + 2b < \infty.$$

Therefore U is a mapping from $\mathcal{W}$ into $\mathcal{W}$. If W and W' belong to $\mathcal{W}$, we have

$$|||UW - UW'|||^2 = E\left\{\sup_{u<t<v}\left\|\int_{]0,t]}(a(W,s)-a(W',s))\,dZ_s\right\|^2\right\}$$

$$\leqslant E\left\{Q_{v^-}\int_{]u,v[}\|a(W,t)-a(W',t)\|^2\,dQ_t\right\}$$

$$\leqslant E\left\{Q_{v^-}\int_{]u,v[}\left[\sup_{s<t}\|W_s - W_s'\|^2\right]L_t\,dQ_t\right\}$$

$$\leqslant \tfrac{1}{2}|||W - W'|||^2.$$

Then U is a contraction on the complete metric space $\mathcal{W}$ with the metric $|||\mathrm{W}-\mathrm{W}'|||$.

According to the most classical fixed point theorem, there exists a unique element X' of E such that $UX' = X'$, which proves the lemma.

Remark and Proposition Now we remark that it is easy to prove Theorem 6.10 in the uniformly lipschitzian case. More precisely: *We consider the hypothesis and notations given in the theorem of Section* 6.10, *except for the hypothesis* (L_2), *which is replaced by* (L_1), *and we assume that* $\int_{]0,t]} L_s\,dQ_s$ *is finite* (*for all* $(\omega, t) \in \Omega \times T$). *Then there exists a strong solution of* (6.1.1) *on* $T = [0, t_m]$.

Indeed, let us define

$$A_t := Q_t \int_{]0,t]} \|a(0,s)\|^2 \, dQ_s.$$

We define recursivsly the sequence $(X^k)_{k>0}$ of strong solutions of (6.1.1) and the associated sequence $(u(k))_{k\geq 0}$ of stopping times as follows:

$u(0) = 0$;

$b(k)$ is a positive number such that

$$P\left(\left[X^*_{u(k)} + A_{t_m} + V^*_{t_m} > b(k)\right]\right) \leq 2^{-k};$$

if $X^*_{u(k)} > b(k)$, we set

$$u(k+1) := u(k);$$

if $X^*_{u(k)} \leq b(k)$, we set

$$u(k+1) := \inf\left\{ t : t \geq u(k), A_t + V^*_t > b(k) \right.$$

$$\left. \text{or} \quad Q_t \int_{]u(k),t]} L_s \, dQ_s > \tfrac{1}{2} \right\} \wedge t_m.$$

X^{k+1} is the strong solution of (6.1.1) defined up to P-equivalence on $[0, u(k+1)]$: this solution exists according to the previous lemma. It is easily seen that

$$\lim_{k \uparrow \infty} \left[u(k) < t_m \right] = 0,$$

and we can define the strong solution X on T by $X 1_{[0,u(k)]} = X^k 1_{[0,u(k)]}$, which proves the remark.

This result will not be used in the sequel of the proof of the main theorem.

6.13 Extension Lemma: General Situation

We come back to the general hypothesis of the theorem of Section 6.10.

Lemma *We consider the hypothesis and notations given in the theorem of Section* 6.10. *Let u be a stopping time and X a strong solution* (*cf. Section* 6.5) *of* (6.1.1) *on* $[0, u]$. *Then for every* $\epsilon > 0$ *there exists a stopping time* v *with* $P[v > u] \geq P[u < t_m] - \epsilon$ *and a strong solution* X' *of* (6.1.1) *on* $[0, v]$.

Proof Of course (see Section 6.11) the processes $X 1_{[0,u]}$ and $X' 1_{[0,u]}$ are P-equivalent. Let $b > 0$ be such that $P([X^*_u + A_{t_m} + V^*_{t_m} > \frac{1}{2} b]) \leq \frac{1}{2}\epsilon$ and L_b be the associated process occurring in condition (L_2). Let v' be the stopping time defined as follows:

if $X_u^* > \frac{1}{2}b$,

$v' := u$;

if $X_u^* \leqslant \frac{1}{2}b$,

$$v' := \inf\left\{ t : t > u,\ A_t^* + V_t^* > \tfrac{1}{2}b \quad \text{or} \quad Q_t \int_{]u,\,t]} L_b(s)\, dQ_s > \tfrac{1}{2} \right\} \wedge t_m.$$

For every **H**-valued cadlag function f defined on T, we put

$$f^b(t) := f(t)\left[1 \wedge (b/\|f(t)\|)\right]$$

and

$$a^b(f, \omega, t) = a(f^b, \omega, t)1_{]u,\,t_m]} + a(f, \omega, t)1_{[0,\,u]}.$$

The process f^b is clearly adapted. Therefore, following the proposition in Section 6.4, $a^b(f, \omega, t) = a(f^b, \omega, t)$ is predictable. Then, using the easily verified inequality

$$\left\| f(s)\left(1 \wedge \frac{b}{\|f(s)\|}\right) - g(s)\left(1 \wedge \frac{b}{\|g(s)\|}\right) \right\| \leqslant 2(\|f(s) - g(s)\|),$$

we obtain

$$\|a^b(f, \omega, t) - a^b(g, \omega, t)\| \leqslant 2L_b(\omega, t) \sup_{s<t} \|f(s) - g(s)\| \text{ on }]u, v'[$$

for every f and g. Then we can apply the lemma of Section 6.12 with the functional a^b, and there exists a strong solution X' on the stochastic interval $[0, v']$ of the equation

$$X_t' = V_t + \int_{]0,\,t]} a^b(X, s)\, dZ_s.$$

Let v be the stopping time defined by

$$v := \inf\{ t : t \geqslant u,\ t \leqslant v',\ \|X_t'\| > b \} \wedge t_m.$$

The right continuity of X' and the definition of v' show that $P[v > u] \geqslant P[u < t_m] - \epsilon$. But $a^b(X, s)1_{]u,\,v]} = a(X, s)1_{]u,\,v]}$ and X' is a strong solution of (6.1.1) on the stochastic interval $[0, v]$, which proves the lemma.

6.14 Maximal Solutions

Now, we complete the proof of the theorem of Section 6.10 by showing that there exists one stopping time v and a solution X of (6.1.1) on $[0, v]$ such that (i) holds and this v is predictable.

We consider the family S of stopping times u such that there exists a strong solution X of (6.1.1) on the stochastic interval $[0, u]$. We denote by v

the essential supremum (for the probability P) of those stopping times u that belong to S.

If u and u' are two elements of S and X and X' the associated solutions, then

$$X'' := X1_{[0,u]} + X'1_{]u,u\vee u']}$$

is a solution of (6.1.1) on the stochastic interval $[0, u \vee u']$ (see Section 6.11), and $(u \vee u')$ belongs to S. Then there exists an increasing sequence $(u(n))_{n>0}$ of elements of S such that $v = \lim_n u(n)$ a.s. For every integer n, let X^n be a strong solution of (6.1.1) on $[0, u(n)]$; we define the process X by $X1_{[0,u(n)]} = X^n 1_{[0,u(n)]}$ (see Section 6.11); X is a strong solution of (6.1.1) on $[0, v[$.

We consider now the following family of stopping times

$$w(n,k) := u(n) \wedge \inf\{t : \|X_t\| > k\},$$

$$w(k) := \sup_n w(n,k).$$

We have $w(k) \leqslant v$, and we prove that

$$P([w(k) = v < t_m]) = 0. \tag{6.14.1}$$

On the contrary, let us assume that there exists an integer k such that

$$P([w(k) = v < t_m]) = 2\epsilon > 0.$$

Since the process X is bounded by k on $[0, w(k)[$, there exists a strong solution X' of (6.1.1) on $[0, w(k)]$ (see Section 6.5). Now according to Lemma 6.13, there exists a stopping time w' which belongs to S and such that

$$P([w' > w(k)]) \geqslant P[w(k) < t_m] - \epsilon.$$

This implies that

$$P([w' > v]) \geqslant \epsilon,$$

which contradicts the definition of v and proves (6.14.1). But the definition of $w(k)$ shows that $v = \lim_k w(k)$. This equality and (6.14.1) then show the predictability of v.

6.15 Comments and Extensions

1 *On the functional a.* In the presentation of this chapter we have assumed that a is defined on the set $\mathbf{D} \times \Omega \times T$. We could have defined a bluntly as an adapted (in an obvious sense) mapping from the set of cadlag $\mathbf{H}$-valued processes into the set of locally bounded predictable $\mathbf{L}$-valued processes; the Lipschitz conditions (L_1) and (L_2) would then read

(L_1) $\|(aX - aX')_t\| \leqslant L_t(X - X')^*_{t-}$ for every $t \in T$, X, X' cadlag;

(L_2) $\|(aX - aX')_t\| \leqslant L_b(t)(X - X')^*_{t-}$ for every $t \in T$, X, X' cadlag such that $X^*_t \leqslant b$, $X'^*_t \leqslant b$.

(We recall that for every process Y, Y^*_t stands for $\sup_{s \leqslant t}\|Y\|_s$.) Nothing would have to be changed in the proof, and the same existence and unicity result holds.

In [Pro-1, 2, 3] and [Eme-1, 2], the functional a maps cadlag processes into left continuous ones.

2 *On the stochastic driving term.* Semimartingales as stochastic driving terms have been considered by Protter, Emery, Doléans-Dade, and Meyer. Since in the real case semimartingales and π^*-processes are the same, the theory presented here is also an account of their results. The fundamental role of the π^*-domination property is underlined by the fact that it works for Hilbert-valued processes and that in this case the class of interesting driving terms is wider than the class of semimartingales for a very simple example. (See Section 10.10, Exercise 4.)

3 *More on the stochastic integral with respect to π^*-processes and extensions.* Let Z be a π^*-process $*$-dominated by Q. Since the process Z is a π-process, the stochastic integral $\int Y\,dZ$ is defined for any locally bounded predictable process (see Section 2). Actually, in this setting we can say more; indeed, the stochastic integral $\int Y\,dZ$ is defined for any predictable process Y such that for all $t \in T$

$$S_t < +\infty \text{ a.s.},$$

where

$$S_t := \int_{]0,\,t]} \|Y_s\|^2\,dQ_s$$

(consider the stopping time $u(n)$ defined by $u(n) := \inf\{t : Q_t + S_t > n\}$; the process $\int Y\,dZ$ can be defined on $[0, u(n)[$ exactly as in Section 2; then $\int Y\,dZ$ is defined on $\Omega \times T$ as in Section 2).

Let us consider more generally a locally convex vector space $\mathbf{K}$ and a separable Banach space $\mathbf{H}$. We denote by Z a given $\mathbf{K}$-valued adapted process, and we are given also a vector space $\mathbf{L}$ of linear operators from $\mathfrak{L}^0_{\mathbf{K}}(\Omega, \mathcal{F}, P)$ into $L^0_{\mathbf{H}}(\Omega, \mathcal{F}, P)$.

If, as in Section 2.2, we call $\mathcal{E}(\mathbf{L})$ the set of step processes of the form

$$Y := \sum_{i=1}^{n} u_i 1_{F_i \times]s_i,\, t_i]},$$

where $u_i \in \mathbf{L}$ and the $F_i \times]s_i, t_i]$ are predictable rectangles, we can define the stochastic integral $\int Y\,dZ$ since

$$\sum_{i=1}^{n}\left[u_i(Z_{t_i})-u_i(Z_{s_i})\right]1_{F_i}\in L^0_{\mathbf{H}}(\Omega,\mathcal{F},P).$$

We shall make the following assumptions on $\mathbf{L}$ and Z:

(a) for every $u\in\mathbf{L}$ the process $u(Z)$ has a cadlag adapted version. We may therefore speak of the cadlag integral process $(\int Y\,dZ):=(\int 1_{]0,t]}Y\,dZ)_{t\in T}$ as in Section 2 for every $Y\in\mathcal{E}(\mathbf{L})$;

(b) $\mathbf{L}$ is equipped with a norm $\|\cdot\|$, giving it the structure of Banach space;

(c) there exists an increasing positive finite process Q such that the following inequality holds for every $Y\in\mathcal{E}(\mathbf{L})$ and every stopping time u:

$$E\left\{\sup_{0\leqslant t<u}\left\|\int_{]0,t]}Y_s\,dZ_s\right\|^2\right\}\leqslant E\left\{Q_{u-}\int_{]0,u[}\|Y_s\|^2\,dQ_s\right\}. \qquad (*)$$

As an immediate consequence of property (c) the stochastic integral process $(\int Y\,dQ)$ can be defined for every process Y with the property $\int_{]0,t]}\|Y_s\|^2\,dQ_s<\infty$: the random variables $\int_{]0,t]}Y\,dZ$, $t\in T$, are defined by continuity for processes Y such that $E\{Q_t\int_{]0,t]}\|Y_s\|^2\,dQ_s\}<\infty$; the uniform approximation theorem holds as in Section 2.5, and the definition of the cadlag process $(\int Y\,dZ)$ follows exactly, word for word, as in Section 2. Then using the localization procedure described in **3** above, we extend the definition from processes Y such that $E\{Q_t\int_{]0,t]}\|Y_s\|^2\,dQ_s\}<\infty$ for all t to processes Y such that $\int_{]0,t]}\|Y_s\|^2\,dQ_s<\infty$ a.s. for all t.

4 *A generalized version of Theorem* 6.10. Remarks **1–3** lead to the following abstract generalized form of stochastic equations with Lipschitz conditions.

We suppose that we are given Z and $\mathbf{L}$ as in **3** above and a functional $\tilde{a}$ which to every cadlag $\mathbf{H}$-valued adapted process associates a locally bounded predictable process $\tilde{a}(X)$, which is *adapted* in the following sense: $X_s=X'_s$ a.s. for all $s<t$ implies $[\tilde{a}(X)]_s=[\tilde{a}(X')]_s$ on $[0,t]$ up to P-equivalence.

For the functional $\tilde{a}$ we make the following assumptions, which is a weakening of the Lipschitz condition (L_2):

(L'_2) for every $b>0$ there exists a cadlag adapted increasing process $\bar{Q}^b$ such that for any two cadlag $\mathbf{H}$-valued processes X and X' bounded by b and any $s<t$

$$\int_{]s,t]}\|\tilde{a}(X)-\tilde{a}(X')\|^2\,dQ\leqslant\int_{]s,t]}\|(X'-X)^*_r\|^2\,d\bar{Q}^b_r,$$

where for every process Y, $Y^*_t:=\sup_{s\leqslant t}\|Y_s\|$.

We have then the "abstract" form of the theorem of Section 6.10:

Theorem *Let* **H**, **K**, **L**, Z, $\tilde{a}$ *be as just defined with the assumed properties* (a), (b), (c), *and* (L_2'). *Let* V *be an* **H**-*valued cadlag adapted process. Then there exists a unique stopping time* v *and an* **H**-*valued cadlag adapted process* X, *defined on* $]0, v[$, *with the following properties*:

(i) *if* $\omega \in \{v < t_m\}$, $\limsup_{t \uparrow v(\omega)} \|X_t(\omega)\| = +\infty$;
(ii) $X_t := V_t + \int_{]0,t]} \tilde{a}(X)\, dZ_s$ *on* $]0, v[$.

The stopping time v *is actually predictable.*

Proof The proof is a mere rephrasing of the successive steps leading to Theorem 6.10.

7 PROPERTIES OF SOLUTIONS; CONDITIONS FOR NONEXPLOSION AND STABILITY

Now we give some conditions for nonexplosion, and we study the stability of the solutions.

The simplicity of the proofs given below exemplifies the usefulness of the inequality introduced with π^*-processes.

7.1 A Lemma of Gronwall Type

Lemma *Let* A *be an adapted increasing process defined on the stochastic interval* $[0, u[$, *and for every* $l > 0$ *let us denote by* A^l *the process* $A \wedge l$. *Then for every adapted increasing process* ϕ *satisfying*

$$E(\phi_{v^-}) \leqslant K + \rho E\left\{\int_{[0,v[} \phi_{s^-}\, dA_s^l\right\}$$

for all stopping time $v \leqslant u$, *the following inequality holds*:

$$E(\phi_{u^-}) \leqslant R(l, K, \rho) := 2K \sum_{j=0}^{[2\rho l]} (2\rho l)^j,$$

where $[x]$ *denotes the integer part of* $x \geqslant 0$.

Proof Let us define recursively the following increasing sequence of stopping times:

$$v_0 := 0$$
$$\vdots$$
$$v(k+1) := \inf\{t : A_t^l - A_{v(k)}^l \geqslant 1/(2\rho)\} \wedge u.$$

Writing

$$x_k := E(\phi_{v(k)^-}),$$

and taking into account that $A^l_{u-} \leqslant l$, we may write

$$\begin{aligned} x_{k+1} &\leqslant K + \rho E\left\{\int_{[0,\,v(k)]} \phi_{s-}\, dA^l_s\right\} + \rho E\left\{\int_{]v(k),\,v(k+1)[} \phi_{s-}\, dA^l_s\right\} \\ &\leqslant K + \rho l x_k + \tfrac{1}{2} E(\phi_{v(k+1)-}) \\ &\leqslant K + \rho l x_k + \tfrac{1}{2} x_{k+1}. \end{aligned}$$

From this we derive

$$x_{k+1} \leqslant 2K + 2\rho l x_k,$$

and

$$x_{k+1} \leqslant 2K \sum_{j=0}^{k} (2\rho l)^j \qquad \text{for all} \quad k.$$

Since A^l is bounded by l on $[0, u[$, we have $v(k) = u$ as soon as $k \geqslant 2\rho l$. Therefore we derive immediately the conclusion of the lemma from the above inequality.

7.2 Nonexplosion

Theorem *We consider the hypothesis and notations given in the theorem of Section* 6.10. *Moreover, we suppose that the following condition is fulfilled*:

[i] *there exists $d > 0$ such that for every element (f, ω, t) of $(D^{\mathbf{H}} \times \Omega \times T)$ we have*

$$\|a(f, \omega, t)\|^2 \leqslant d\Big(1 + \sup_{s<t} \|f(s)\|^2\Big).$$

Then, there is a strong solution (*see Section* 6.5) *of* (6.1.1) *on* $[0, t_m]$.

Proof Let (X, u) be a maximal solution, i.e., a solution in $[0, u[$ with the property [i] of the theorem of Section 6.10. We have to prove that $\lim_{t\uparrow u,\, t<u} \|X_t\| < \infty$ a.s. At the same time, this will imply that $u \geqslant t_m$ a.s. and (cf. the remark in Section 6.5) the existence of a solution X on $[0, t_m]$.

We consider the following increasing sequence (u_n) of stopping times:

$$u_n := \inf\{t : t \in [0, t_m],\, Q_t + V^*_t > n\} \wedge u.$$

We have clearly

$$\lim_n P\{u_n < u\} = 0. \tag{7.2.1}$$

From the definition of a solution X and the fact that Z is a π^*-process we may write for every stopping time $v \leqslant u_n$

$$E(X^*_{v-})^2 \leqslant 2E(V^*_{v-})^2 + 2E\left\{Q_{v-} \int_{]0,\,v[} \|a(X, s)\|^2\, dQ_s\right\}.$$

The hypothesis of the theorem shows that

$$E(X^*_{v-})^2 \leqslant 2n^2(1+d) + 2dn^2 \int_{]0,\,v[} (X^*_{s-})^2\, dQ_s.$$

We may therefore apply the lemma of Section 7.1 with $\phi = (X^*)^2$, which gives in particular

$$E\left(X^*_{u_n^-}\right)^2 < \infty,$$

and therefore

$$X^*_{u_n^-} < \infty \text{ a.s.}$$

The theorem follows then from (7.2.1) and the remarks at the beginning of the proof.

7.3 A Topology on π^*-Processes (Notation F)

For every π^*-process Z defined on $[0, t_m]$, we set

$$F(Z) := \inf E\left(1 \wedge Q_{t_m}\right),$$

where this infimum is taken for all processes Q that $*$-dominate Z (see Section 6.7).

The function F clearly takes a finite positive value for every π^*-process Z and possesses the property

$$F(Z + Z') \leqslant 2F(Z) + 2F(Z').$$

This inequality follows from the fact that if Q $*$-dominates Z and Q' $*$-dominates Z', $2(Q + Q')$ $*$-dominates $Z + Z'$.

Since $\lim_{\lambda\to 0} F(\lambda Z) = 0$, we define a topology of topological vector space on the vector spaces of $\mathbf{K}$-valued π^*-processes by taking the following basis of neighborhoods of 0:

$$\{Z : F(Z) \leqslant \epsilon\},\ \epsilon \in \mathbf{R}^+ - \{0\}.$$

This is a metrizable topology since it has a denumerable basis of neighborhoods of 0. We will denote by $\Pi^*_{\mathbf{K}}$ the space of $\mathbf{K}$-valued π^*-processes endowed with this topology.

7.4 The Space Π^* is Complete

Lemma *Let $(Z^n)_{n\geqslant 0}$ be a sequence of π^*-processes. For every n we consider an increasing process Q^n which $*$-dominates $Z^{n+1} - Z^n$. If we assume for every n that $E(1 \wedge Q^n_{t_m}) \leqslant 8^{-n}$, the sequence Z^n converges uniformly on $[0, t_m]$ for P-almost all paths to a process Z $*$-dominated by the finite increasing process $Q := \sum_{n\geqslant 0} 2^n Q^n$, while $Z - Z^n$ is $*$-dominated by $\sum_{k\geqslant 0} 2^{k+1} Q^{n+k}$ for all $n \geqslant 0$.*

Proof Let us first remark that $P\{Q^n_{t_m} > 4^{-n}\} \leqslant 2^{-n}$ for every n. The Borel–Cantelli lemma therefore shows that $Q_{t_m} < \infty$ a.s. Since the processes Q^n are positive and increasing, the convergence of the series $\sum_{n \geqslant 0} 2^{n+1} Q^n$ is therefore uniform on every path a.s. Introducing next the sets

$$G_n := \left\{ \sup_{t < t_m} \|Z^n_t - Z^{n+1}_t\|^2 > 2^{-n} \right\}$$

and the stopping time

$$u := \inf\{t : Q^n_t \geqslant 1\} \wedge t_m,$$

$$2^{-n} P(G_n) \leqslant 2^{-n} P(G_n \cap \{u \geqslant t_m\}) + P\{u < t_m\}.$$

We note that $P\{u < t_m\} \leqslant 8^n$ and on $G_n \cap \{u \geqslant t_m\}$ the inequality $\sup_{t<u} \|Z^n_t - Z^{n+1}_t\|^2 > 2^{-n}$ holds. We may therefore write, remembering that $(Q^n_{u-})^2 < 1 \wedge Q^n_{t_m}$,

$$2^{-n} P(G_n) \leqslant E\left(\sup_{t<u} \|Z^{n+1}_t - Z^n_t\|^2 \right) + 8^{-n}$$

$$\leqslant E(Q^n_{u-})^2 + 8^{-n} \leqslant 4^{-n}.$$

The Borel–Cantelli lemma again shows the uniform convergence of the sequence (Z^n) on $[0, t_m]$ for P-almost all $\omega \in \Omega$ to a cadlag process Z.

From this uniform convergence, for every n, every $\mathscr{Q}$-simple process Y, and every stopping time u we may write

$$E\left(\sup_{0<t<u} \left\| \int_{]0,t]} Y \, d(Z - Z^n) \right\|^2 \right)$$

$$\leqslant \liminf_k E\left(\sup_{0<t<u} \left\| \int_{]0,t]} Y \, d(Z^{n+k} - Z^n) \right\|^2 \right)$$

$$\leqslant \liminf_k E\left(\sum_{r=0}^{k} 2^{r+1} Q^{n+r}_{u-} \int_{]0,u[} \|Y_s\|^2 \, dQ^{n+r}_s \right)$$

$$\leqslant E\left(\tilde{Q}^n_{u-} \int_{]0,u[} \|Y_s\|^2 \, d\tilde{Q}^n_s \right)$$

if we call $\tilde{Q}^n$ the process

$$\tilde{Q}^n := \sum_{k \geqslant 0} 2^{k+1} Q^{n+k}.$$

The lemma follows immediately from the last inequality. (Note that $Q = \tilde{Q}^0$.)

Proposition *Let* **K** *be a Banach space and* $(Z_n)_{n>0}$ *be a Cauchy sequence for F of* **K**-*valued* π^*-*processes, i.e., a sequence such that*

$$\lim_{n\to\infty} \sup_{k>0} F(Z_{n+k} - Z_k) = 0.$$

Then there exists a **K**-*valued* π^*-*process* Z *such that*

$$\lim_{n\to\infty} F(Z - Z_n) = 0.$$

Proof According to the inequality

$$F(Z - Z_j) \leqslant 2F(Z - Z_k) + 2F(Z_k - Z_j),$$

it is sufficent to prove the existence of a convergent subsequence; thus we may suppose first that for every integer n

$$F(Z_n - Z_{n+1}) \leqslant 8^{-n}.$$

In this case let us consider for every n a process Q^n which *-dominates the process $(Z_n - Z_{n+1})$ with $E(1 \wedge Q^n_{t_m}) \leqslant 8^{-n}$; we may apply the previous lemma and obtain the proposition.

Remark When in Section 10 we prove that semimartingales are π^*-processes, we shall compare the topology just defined with topologies considered by Protter and Emery in stability problems analogous to those studied here.

7.5 Stability of the Solutions of Equation (6.1.1)

Our purpose here is to show that a "small" change in Z in the functional a and in the process V produces only a "small" perturbation on the paths of the solution process X.

More precisely, we consider two π^*-processes Z and Z', two **H**-valued cadlag adapted processes V and V', and two $\mathfrak{L}(\mathbf{K};\mathbf{H})$-valued processes a and a' assumed to be predictable with respect to the canonical basis $\mathbf{B}^{\mathbf{H}}$ (see Section 6.2).

We assume that there exists an increasing adapted process L such that for all $f \in D^{\mathbf{H}}$, $g \in D^{\mathbf{H}}$ we have

$$\|a(f,t) - a(g,t)\| \leqslant L_t \sup_{s<t} \|f(s) - g(s)\| \tag{7.5.1}$$

and

$$\|a'(f,t) - a'(g,t)\| \leqslant L_t \sup_{s<t} \|f(s) - g(s)\| \tag{7.5.2}$$

(Lipschitz property (L_1)).

We denote by X and X' the strong solutions of the two stochastic differential equations

$$X_t = V_t + \int_{]0,t]} a(X,s)\, dZ_s, \tag{7.5.3}$$

$$X'_t = V'_t + \int_{]0,t]} a'(X',s)\, dZ'_s, \tag{7.5.4}$$

and we assume these solutions to be defined on $[0, u[$, where u is a stopping time. We consider equation (7.5.4) to be the "perturbed" equation. The smallness of the perturbation is measured through the parameters

$$d := E\left\{ \sup_{t<u} \|a(X,t) - a'(X,t)\|^2 \right\}, \tag{7.5.5}$$

$$d' := E\left\{ \sup_{t<u} \| V_t - V_t'\|^2 \right\}, \tag{7.5.6}$$

$$c := E\left\{ \sup_{t<u} \|a(X,t)\|^2 \right\}. \tag{7.5.7}$$

We assume these numbers d, d', and c to be finite. We shall denote by A a process which *-dominates Z. Then we can state the following result:

Theorem *Let Q be a process which *-dominates $Z - Z'$. Under the above assumptions, for every $\epsilon > 0$ there exists a stopping time $u_\epsilon \leqslant u$ with the properties*

(i) $P\{u_\epsilon < u\} \leqslant \epsilon$,
(ii) $E(\sup_{t<u_\epsilon} \|X_t - X_t'\|^2) \leqslant R_\epsilon(d, d', q)$,

where q is such that $P\{Q_{u^-} \geqslant q\} \leqslant \epsilon/2$, and the function R_ϵ is determined only by the process A, L and the functional a and is such that

$$\lim_{d', d, q \to 0} R_\epsilon(d, d', q) = 0.$$

Proof We define the following stopping times:

$$w^l := \inf\{t : A_t \vee L_t > l\} \wedge u,$$

$$w_q' := \inf\{t : Q_t > q\} \wedge u,$$

$$u_\epsilon := w^l \wedge w_q'.$$

We can clearly choose l in such a way that $P\{w^l < u\} \leqslant \epsilon/2$ and can choose q in such a way that $P\{Q_{u^-} \geqslant q\} \leqslant \epsilon/2$, and therefore $P\{w_q' < u\} \leqslant \epsilon/2$. Hence we get $P\{u_\epsilon < u\} \leqslant \epsilon$. From the definition of X and X' we have

$$\begin{aligned} X_t - X_t' = V_t - V_t' &+ \int_{]0,t]} (a(X,s) - a'(X,s))\, dZ_s \\ &+ \int_{]0,t]} (a'(X,s) - a'(X',s))\, dZ_s \\ &+ \int_{]0,t]} (a'(X',s) - a'(X,s))\, d(Z_s - Z_s') \\ &+ \int_{]0,t]} (a'(X,s) - a(X,s))\, d(Z_s - Z_s') \\ &+ \int_{]0,t]} a(X,s)\, d(Z_s - Z_s'). \end{aligned}$$

From the property of π^*-processes we derive for every stopping time $v \leqslant u$

$$E\Big(\sup_{t<v} \|X_t - X_t'\|^2\Big) \leqslant 6E\Big(\sup_{t<v} \|V_t - V_t'\|^2\Big)$$
$$+ 6E\Big(A_{v^-} \int_{]0,v[} \|a(X,s) - a'(X,s)\|^2 \, dA_s\Big)$$
$$+ 6E\Big(A_{v^-} \int_{]0,v[} \|a'(X,s) - a'(X',s)\|^2 \, dA_s\Big)$$
$$+ 6E\Big(Q_{v^-} \int_{]0,v[} \big(\|a'(X',s) - a'(X,s)\|^2$$
$$+ \|a'(X,s) - a(X,s)\|^2 + \|a(X,s)\|^2\big) \, dQ_s\Big).$$

If we set

$$\phi_s = \sup_{t<s} \|X_t - X_t'\|^2,$$

for every stopping time $v \leqslant u_\epsilon$ clearly we get

$$E(\phi_{v^-}) \leqslant 6(d' + dl^2 + dq^2 + cq^2) + 6l^2(l+q)E\Big(\int_{]0,v[} \phi_s(dA_s^l + dQ_s^l)\Big),$$

denoting, as in the lemma of Section 7.1, the processes $A \wedge l$ and $Q \wedge l$ by A^l and Q^l, respectively. If we set

$$K := 6(d' + dl^2 + dq^2 + cq^2) \qquad \text{and} \qquad \rho := 6l^2(l+q)$$

and notice that

$$A_s^l + Q_s^l \leqslant (A_s + Q_s) \wedge 2l,$$

The lemma of Section 7.1 shows that the function

$$R_\epsilon(d, d', q) = 2K \sum_{j=0}^{[4\rho l]} (4\rho l)^j$$

fulfills all the requirements of the theorem.

Corollary 1 *Let equation* (7.5.3) *satisfy all the hypotheses of the existence and uniqueness theorem of Section* 6.10, *and let X be a solution of this equation on the stochastic interval* $[0, u[$. *Then equation* (7.5.4) *admits a solution X' on* $[0, u[$ *and the conclusions of the theorem of Section* 7.5 *hold for* $\|X - X'\|$.

Proof Let v be any stopping time $v \leqslant u$ such that (7.5.4) admits a solution on $[0, v[$. Such stopping times exist according to the extension lemma of Section 6.12. Considering then the solution X of (7.5.3) and X' of (7.5.4) on the stochastic interval $[0, v[$, we may apply the above theorem which shows that with a probability as small as we like there is no explosion for X'

before v as long as there is no explosion for X before v. Therefore the maximal solution (X', v_m) for (7.5.4) is such that $v_m \geqslant u$. This proves Corollary 1.

Corollary 2 *Let equation* (7.5.3) *satisfy all the hypotheses of the existence and uniqueness theorem of Section* 6.10 *and let X be a solution of this equation on the stochastic interval* $[0, u[$. *We consider a sequence of equations*

$$X_t^n = V_t^n + \int_{]0,t]} a^n(X^n, s)\, dZ_s^n \qquad (7.5.8)$$

fulfilling the hypotheses of the theorem with a^n admitting the same Lipschitz coefficient process L as a (cf. inequality (7.5.1). *We set*

$$d_n := E\Big\{ \sup_{t<u} \|a(X,t) - a^n(X,t)\|^2 \Big\},$$

$$d_n' := E\Big\{ \sup_{t<u} \|V_t - V_t^n\|^2 \Big\},$$

$$\lim_{n\to\infty} d_n = \lim_{n\to\infty} d_n' = \lim_{n\to\infty} F(Z - Z^n) = 0.$$

Then

(1) *as soon as $d_n < \infty$, $d_n' < \infty$, equation* (7.5.8) *has a unique solution on* $[0, u[$.

(2) *the random variables* $\sup_{t<u}\|X_t - X_t^n\|^2$ *converge in probability to zero.*

Proof The first part of Corollary 2 is already expressed in Corollary 1. To prove the second part we are left to show that for every $\epsilon > 0$ there exists an integer n_ϵ and for every $n > n_\epsilon$, a stopping time $u_\epsilon^n < u$ such that

$$P\{u_\epsilon^n < u\} \leqslant \epsilon \qquad \text{and} \qquad E\Big\{ \sup_{t<u_\epsilon^n} \|X_t - X_t^n\|^2 \Big\} \leqslant \epsilon.$$

Let us denote by Q^n a cadlag increasing adapted process which *-dominates $Z - Z^n$ and such that

$$q_n^2 := E(1 \wedge Q_{t_m}^n) \leqslant 2F(Z - Z_n).$$

We choose n_ϵ big enough to get $q_n < \epsilon/2 < 1$ for every $n \geqslant n_\epsilon$. With this condition $P\{Q_{t_m}^n > q_n\} \leqslant q_n < \epsilon/2$ for all $n > n_\epsilon$. As in the proof of the theorem, we choose l depending only on A, L, and ϵ such that $P\{A_{t_m} \vee L_{t_m} \geqslant l\} \leqslant \epsilon/2$, and we define the stopping time u_ϵ^n as the stopping time u_ϵ of the theorem; we obtain

$$P\{u_\epsilon^n < u\} \leqslant \epsilon \qquad \text{for all} \quad n > n_\epsilon$$

and

$$E\left(\sup_{t<u_\epsilon^n}\|X_t - X_t^n\|^2\right) \leqslant 2K\sum_{j=0}^{[4\rho l]}(4\rho l)^j \qquad \text{for all} \quad n > n_\epsilon,$$

where

$$K = 6\left(d_n + d_n l^2 + d_n q_n^2 + cq_n^2\right),$$
$$\rho = 6l(l + q_n).$$

Since c is fixed and l depends only on ϵ, it is then clear that we can choose $n'_\epsilon > n_\epsilon$ so that we have at the same time

$$P\{u_\epsilon^n < u\} \leqslant \epsilon \qquad \text{for all} \quad n > n'_\epsilon$$

and

$$E\left(\sup_{t<u_\epsilon^n}\|X_t - X_t^n\|^2\right) \leqslant \epsilon \qquad \text{for all} \quad n > n'_\epsilon.$$

Thus Corollary 2 is proved.

EXERCISES

1 Let $(\Omega, \mathfrak{F}, P, (\mathfrak{F}_t)_{t\in T})$ be a stochastic basis with $T := [0, 1]$. Let $(Z_t)_{t\in T}$ and $(V_t)_{t\in T}$ be two real continuous π-processes. We put

$$W_t := -V_t + \tfrac{1}{2}[V, V]_t;$$
$$C_t := \exp(-W_t)\int_0^t(\exp(W_s))\,(dZ_s - d[Z, V]_s).$$

Compare the process C and the process Y, which is a solution of the following differential equation:

$$Y_t := Z_t - Z_0 + \int_0^t Y_s\,dV_s.$$

[*Hint*: You can apply the Ito formula to the function $F(a, b) := be^{-a}$ and to the processes (A, B), where $A_t := W_t$ and $B_t := \int_0^t(\exp(W_s))\,(dZ_s - d[Z, V]_s)$, and use some results in Section 4.]

2 Let f and g be two real functions defined on the real line; we suppose that the derivatives f' and g' of f and g are continuous; moreover, we suppose that for each real number x we have $g'(x) = f[g(x)]$. We put $T := [0, 1]$.

(1) Let $(X_t)_{t\in T}$ be a continuous adapted π-process (see Section 3); let $A := [X, X]$ be the quadratic variation of X. We consider the differential equation

$$Y_t = g(0) + \int_0^t f(Y_s)\,[dM_s + dA_s].$$

Show that the solution of this differential equation can be written $Y_t = g(Z_t)$, where $Z_t = X_t + A_t - \frac{1}{2}\int_0^t (f'g(Z_s))\, dU_s$ and U is a continuous adapted process with bounded variation (calculate U). [*Hint*: Apply the Ito formula to $g(Z_t)$ and show that $g(Z_t)$ is a solution of a differential stochastic equation.]

(2) Check the equation $g'(x) = f[g(x)]$ in the following two cases:

(i) $f(x) := x$, $g := Ce^x$, where C is a real number;

(ii) $f(x) := \sin x$, $g(x) := \arcsin[1/(\mathrm{ch}(C - x))]$, where C is a real number and when $x < C$.

HISTORICAL NOTES

Section 5. The starting examples of this section are presented in a way, that closedly follows Nelson in [Nel]. The equations with Lipschitz hypotheses and possibly nonbrownian stochastic driving terms, generalizing the Ito equation, were introduced by Protter [Pro-1] and further studied by N. Kazamaki (see [Kaz]), C. Doleans-Dade and P.A. Meyer (see [Dol-4], [DoM-2]). The reader will find in [Mey-4] the particular equations $dX = X^- \, dZ$ and $dX = \dot{X}\, dZ$ (where $\dot{X}$ is the predictable projection of the cadlag process X, while X^- is the left continuous version) with their explicit solutions.

Section 6. The general stochastic equations considered in this section were introduced at about the same time and, as far as real processes are concerned, in close forms by Emery [Eme-1] and the authors [MeP-7]. The notion of π^*-processes was introduced by the authors and seems to offer the right framework for dealing with real and Hilbert processes at the same time. The material presented here is essentially that of the papers of [MeP-6] and [MeP-7].

Section 7. The stability theorems given here are due to the authors. They generalize to our context of π^*-processes and with our method the results obtained successively by Protter [Pro-3] and Emery [Eme-3]. The topologies, equivalent to the topology considered here, when real semimartingales are considered, were introduced and studied by Meyer, Protter, and Emery.

CHAPTER 4

MARTINGALES AND SEMIMARTINGALES

Martingales, submartingales, and supermartingales were introduced in Section 1.17, where some of their basic properties were then studied. This chapter is devoted to the deeper results of the theory as developed mainly by J. L. Doob, P. A. Meyer, D. Burkholder, C. Dellacherie, and others, with a complementary result on the relations between semimartingales and π^*-processes.

The presentation of the theory adopted here is quite concise, although it contains all the major results, and the treatment is in the infinite-dimensional case.

In Section 8 we study properties related to equi-integrability properties, namely, the existence of a σ-additive Doléans measure and the stopping theorem.

Section 9 is devoted to the Doob–Meyer decomposition of an "admissible measure" and to the decomposition of a local martingale.

In Section 10 we deal with the relations between π^*-processes and semimartingales.

In Section 11 the classical Davis inequalities (with improved constants) are proved. The procedure adopted here is pretty close to the initial Burkholder "transform method." Following Garsia's approach, the Fefferman inequality is proved first.

8 MARTINGALES AND SUBMARTINGALES: EQUI-INTEGRABILITY AND TIED PROPERTIES

In this section we return to the study of the basic properties of martingales and submartingales, which are related to the existence of a σ-additive Doléans measure. The existence of such a measure was proved for positive submartingales in Section 1. We show in Section 8.4 that it is equivalent to an equi-integrability property under completeness assumptions on the σ-algebras $\mathcal{F}_t$.

In this whole section we consider a probabilized stochastic basis $(\Omega, \mathcal{F}, P, (\mathcal{F}_t)_{t\in T})$, with $\mathcal{F}_t = \mathcal{F}_{t^+}$ as we have always assumed; we shall note as before, $t_\infty = \sup\{t : t \in T\}$, the supremum being taken in $\overline{\mathbf{R}}^+$, and write t_m instead of t_∞ when we want to specify $t_\infty < \infty$.

8.1 The Stopping Theorem: General Form

Theorem *Let X be a cadlag process with values in the Banach space $\mathbf{H}$, such that for every element t of T X_t belongs to $L^1_{\mathbf{H}}(\Omega, \mathcal{F}, P)$. We assume that the Doléans function $d(X)$ of X is σ-additive and that for every decreasing sequence $(u(n))_{n>0}$ of simple stopping times the associated sequence $(X_{u(n)})_{n>0}$ of random variables is equi-integrable.*

Let u be a stopping time. Let X' be the process X stopped at u (i.e., $X'_t = X_{t\wedge u}$). Then for every element B of $\mathcal{Q}$ we have

$$[d(X')](B) = [d(X)](B \cap]0,u]).$$

Proof —The property is easily checked when u is a simple stopping time (we leave this verification to the reader). We consider now a general stopping time u and a process X fulfilling the assumptions of the theorem. Let $(u(n))_{n>0}$ be a sequence of simple stopping times which decreases to u (see Section 1.11). Let $B := F \times]s,t]$ be an element of $\mathcal{R}$; we have

$$[d(X')](B) = E[1_F(X_{t\wedge u} - X_{s\wedge u})] = \lim_{n\to\infty} E[1_F(X_{t\wedge u(n)} - X_{s\wedge u(n)})]$$

since the sequence $(X_{t\wedge u(n)})_{n>0}$ is equi-integrable and converges a.s. to $X_{t\wedge u}$. Thus

$$[d(X')](B) = \lim_{n\to\infty} [d(X)](B \cap]0,u(n)]) = [d(X)](B \cap]0,u]).$$

8.2 Doob's Inequalities

Theorem (1) *Let I be a denumerable set included in $[0,t_\infty] \subset \overline{\mathbf{R}}^+$ with $t_\infty \in I$ and $0 \in I$. If X is a real submartingale indexed by I, then for every $\lambda > 0$ the following inequalities hold:*

$$\lambda P\left\{\sup_{t\in I} X_t > \lambda\right\} \leqslant \int_{\{\sup_{t\in I} X_t > \lambda\}} X_{t_\infty}\, dP \leqslant \int X^+_{t_\infty}\, dP \qquad (8.2.1)$$

and

$$\lambda P\left\{\inf_{t\in I} X_t < -\lambda\right\} \leqslant \int_{\{\inf_{t\in I} X_t \geqslant -\lambda\}} X_{t_\infty}\, dP - \int X_0\, dP. \qquad (8.2.2)$$

(2) *Let p and q be two real positive numbers such that $p > 1$ and $(1/p) + (1/q) = 1$. Let I be a denumerable set in $\mathbf{R}^+$ as above. For every positive submartingale $(X_t)_{t \in I}$ we have*

$$E\left[\left(\sup_{t \in I} X_t\right)^p\right] \leqslant q^p E(X_{t_\infty}^p).$$ [1]

(3) *For every cadlag positive submartingale on $[0, t_\infty]$ and every $p > 1$,*

$$E\left[\left(\sup_{t \in I} X_t\right)^p\right] \leqslant \left(\frac{p}{p-1}\right)^p E(X_{t_\infty}^p).$$

Proof (1) By considering an increasing sequence (I_n) of finite subsets of I such that $I = \bigcup_n I_n$, we clearly reduce the situation to the case in which I is finite. Whether t_∞ is finite or not has no importance. We then assume $t_\infty < \infty$ and define

$$u := \inf\{t : t \in I, X_t > \lambda\},$$

where u is a simple stopping time. From the definition of u and $t_\infty < \infty$ follows

$$\left\{\sup_{t \in I} X_t > \lambda\right\} = \{u \leqslant t_\infty\},$$

and therefore

$$\lambda P\left\{\sup_{t \in I} X_t > \lambda\right\} = \lambda P\{u \leqslant t_\infty\} \leqslant \int_{\{u \leqslant t_\infty\}} X_u \, dP. \tag{8.2.3}$$

Since

$$\int_{\{u \leqslant t_\infty\}} X_u \, dP = \int (X_{u \wedge t_\infty} - X_{t_\infty}) dP + \int_{\{u \leqslant t_\infty\}} X_{t_\infty} \, dP,$$

we derive from (8.2.3) and the definition of $d(X)$

$$\lambda P\left\{\sup_{t \in I} X_t > \lambda\right\} \leqslant -d(X)(]u \wedge t_\infty, t_\infty]) + \int_{\{u \leqslant t_\infty\}} X_{t_\infty} \, dP. \tag{8.2.4}$$

If X is a submartingale, $d(X)$ is positive and (8.2.4) immediately implies (8.2.1). By applying the same inequality (8.2.4) to the supermartingale $-X$ and noticing that in this case

$$d(X)(]u \wedge t_\infty, t_\infty]) \leqslant \int (X_{t_\infty} - X_0) dP,$$

we get

$$\lambda P\left\{\inf_{t \in I} X_t > -\lambda\right\} \leqslant \int (X_{t_\infty} - X_0) dP - \int_{\{u \leqslant t_\infty\}} X_{t_\infty} \, dP$$

which gives inequality (8.2.2).

[1] For the case $p = 1$ the reader is referred to [Mey-1]. See also Section 11 of this book.

(2) The second statement of the theorem follows immediately from (8.2.1) and from the lemma of Section 6.8 in Chapter 3.

(3) The last part of the theorem is a straightforward consequence of (2).

8.3 Equi-Integrability Properties of Submartingales

Proposition (1) *Let X be a cadlag real submartingale (resp. real supermartingale, resp. Banach-valued martingale) on $[0, t_m]$. For every decreasing sequence (u_n) of simple stopping times with values in $[0, t_m]$ the family (X_{u_n}) of random variables is equi-integrable.*

(2) *Let $\mathcal{T}^f[0, t_m]$ be the family of simple stopping times with values in $[0, t_m]$. If X is a positive submartingale or a Banach-valued martingale on $[0, t_m]$, the family $\{X_u : u \in \mathcal{T}^f[0, t_m]\}$ of random variables is equi-integrable.*

Proof (1) Since $-X$ is a submartingale if X is a supermartingale and $\|X\|$ is a submartingale as soon as X is a martingale (see Section 1.18), it is clearly sufficient to prove statement (1) for a submartingale. Let us assume that the family $(X_{u_n} : n \in \mathbf{N})$ is not equi-integrable. Then there exists $\epsilon > 0$ and a subsequence $(u'_n)_{n \in \mathbf{N}}$ of (u_n) such that for all n

$$\int_{\{|X_{u'_n}| > n\}} |X_{u'_n}| \, dP \geqslant \epsilon.$$

Noticing from inequalities (8.2.1) and (8.2.2) that

$$P\{|X_{u'_n}| > n\} \leqslant (1/n) E(|X_{t_\infty}| + |X_0|) \qquad \text{for every} \quad n,$$

we can determine $m > n$ such that

$$\int_{\{|X_{u'_n}| > m\}} |X_{u'_n}| \, dP \leqslant \frac{\epsilon}{4} .$$

Then it becomes clear that we can define recursively a decreasing subsequence $(u'_{n(k)})$ with the following two properties for every $k \in \mathbf{N}$:

$$\int_{\{|X_{u'_{n(k+1)}}| > n(k+1)\}} |X_{u'_{n(k)}}| \, dP \leqslant \frac{\epsilon}{4} ; \tag{8.3.1}$$

$$\int_{\{|X_{u'_{n(k)}}| > n(k)\}} |X_{u'_{n(k)}}| \, dP \geqslant \epsilon. \tag{8.3.2}$$

Let us write F_k for the $\{|X_{u'_{n(k)}}| > n(k)\}$. Since

$$A_k := (F_{k+1} \times]0, t_m]) \cap]u'_{n(k+1)}, u'_{n(k)}]$$

belongs to $\mathcal{Q}$ we have

$$|d(X)(A_k)| = \left| \int_{F_{k+1}} (X_{u'_{n(k)}} - X_{u'_{n(k+1)}}) dP \right| \geqslant \frac{3\epsilon}{4}$$

as a consequence of (8.3.1) and (8.3.2). But this contradicts the fact that X is a submartingale since from the A_ks being disjoint, one should have

$$\sum_k d(X)(A_k) \leqslant E(X_{t_m}) - E(X_0) < \infty.$$

This proves the first part of the proposition.

(2) Since $\|X\|$ is a positive submartingale when X is a Banach-valued martingale, we need only consider the case of a positive submartingale. Let u be any simple stopping time. For every $\lambda > 0$ the set $(\{X_u > \lambda\} \times [0, t_m]) \cap]u, t_m]$ is an element of $\mathcal{Q}$ and its measure for the additive Doléans function $d(X)$ is

$$\int_{\{X_u > \lambda\}} (X_{t_m} - X_u) dP \geqslant 0.$$

Therefore we have

$$\int_{\{X_u > \lambda\}} X_u \, dP \leqslant \int_{\{X_u > \lambda\}} X_{t_m} dP. \tag{8.3.3}$$

But from Doob's inequality (8.2.1),

$$P\{X_u > \lambda\} \leqslant 1/\lambda \int X_{t_m} dP, \qquad \text{for every} \quad u \in \mathcal{T}^f[0, t_m]. \tag{8.3.4}$$

For every $\epsilon > 0$ it is therefore possible to choose $\lambda > 0$ such that

$$\int_{\{X_u > \lambda\}} X_u \, dP < \epsilon.$$

This completes the proof of the proposition.

8.4 A σ-Additivity Lemma for Measures on $\mathcal{Q}$

Lemma *We assume $T = [0, t_m] \subset \mathbf{R}$. Let v be a positive function defined on $\mathcal{Q}$ which satisfies the following three properties:*

(i) *for every pair (A, B) of elements of $\mathcal{Q}$,*

$$v(A) \leqslant v(A \cup B) \leqslant v(A) + v(B);$$

(ii) *for every element s of T,*

$$\lim_{t \downarrow s} v(\Omega \times]s, t]) = 0;$$

(iii) *for every increasing sequence $(u(n))_{n>0}$ of simple stopping times such that*

$$\lim_{n \to \infty} P[u(n) < t_m] = 0$$

we have

$$\lim_{n \to \infty} v(]u(n), t_m]) = 0.$$

Then the following property holds:

(iv) *for every sequence* $(A_n)_{n>0}$ *of elements of* $\mathcal{Q}$ *such that*

$$A_n \underset{n\to\infty}{\downarrow} \emptyset$$

we have

$$\lim_{n\to\infty} v(A_n) = 0.$$

Proof Let $(A_n)_{n>0}$ be a sequence of elements of $\mathcal{Q}$ such that

$$A_n \underset{n\to\infty}{\downarrow} \emptyset.$$

We put

$$a = \tfrac{1}{4} \lim_{n\to\infty} v(A_n);$$

we suppose that $a > 0$, and we derive a contradiction.

The proof is in two steps: first, we approximate "from the inside" the sets A_n by sets D_n such that $v(A_n) \leqslant v(D_n) + a$ for all n. Next we prove that the D_n have been chosen in such a way that $\lim_n v(D_n) = 0$.

(1) For each integer n let $(B(n,k))_{1 \leqslant k \leqslant b(n)}$ be a finite partition of A_n such that for every integer k, $B(n,k)$ is an element of $\mathcal{R}$ (cf. Section 1.8). For every pair (n,k) of integers we have

$$B(n,k) = F(n,k) \times]s(n,k), t(n,k)].$$

Let $s'(n,k)$ be an element of T such that

$$s(n,k) < s'(n,k) < t(n,k)$$

and

$$v(]s(n,k), s'(n,k)]) \leqslant a2^{-n}(1/b(n)).$$

For every integer n we set

$$C(n) := \bigcup_{k=1}^{b(n)} (F(n,k) \times]s'(n,k), t(n,k)]);$$

$$\overline{C}(n) := \bigcup_{k=1}^{b(n)} (F(n,k) \times [s'(n,k), t(n,k)]);$$

$$D(n) := \bigcap_{k=1}^{n} C(k);$$

$$\overline{D}(n) := \bigcap_{k=1}^{n} \overline{C}(k);$$

$$S(n) := \bigcup_{k=1}^{b(n)} (\Omega \times]s(n,k), s'(n,k)]).$$

We have

$$A(n) \subset \{S(n) \cup C(n)\}.$$

If we remember that $A(n) \downarrow \varnothing$, we may write

$$A(n) \subset \left\{ D(n) \cup \left\{ \bigcup_{i=1}^{n} S(i) \right\} \right\},$$

and this implies

$$v[A(n)] \leqslant v[D(n)] + a.$$

(2) Moreover, since $\bar{C}(n)$ is contained in $A(n)$ for every n, we have

$$\bar{D}(n) \underset{n\to\infty}{\downarrow} \varnothing.$$

Let $u(n)$ be the simple stopping time for every n, which is the "beginning" of the set $\bar{D}(n)$, i.e.,

$$u(n)(\omega) := \inf\{ t : t \in T, (\omega, t) \in \bar{D}(n) \},$$

$$u(n)(\omega) := \infty, \qquad \text{as usual if the set above is empty.}$$

Since ω is fixed, we have $\bar{D}(n, \omega) \downarrow \varnothing$; then there exists an integer k such that $\bar{D}(n, \omega) = \varnothing$ (property of compact sets); this means that

$$u(k)(\omega) = t_m.$$

We have thus proved that

$$[u(n) < t_m] \underset{n\to\infty}{\downarrow} \varnothing.$$

Then there exists an integer j such that

$$v(]u(n), t_m]) \leqslant a \qquad \text{for} \quad n \geqslant j \qquad \text{(cf. (iii));}$$

and

$$\begin{aligned} v[A(n)] &\leqslant v[D(n)] + a \qquad \text{(cf. (1))} \\ &\leqslant v(]u(n), t_m]) + a \\ &\leqslant 2a \qquad \text{for} \quad n \geqslant j, \end{aligned}$$

and this contradicts the definition of a.

8.5 The Doléans Measure of a Submartingale of Class D

We call $\mathfrak{T}^f[0, t_m]$, as before, the family of all simple stopping times with values in $[0, t_m]$, while $\mathfrak{T}[0, t_m]$ will stand for the family of all stopping times with values in $[0, t_m]$.

We now give the following definition, which was introduced by P. A. Meyer (see [Mey-1]) for his decomposition theorem (see Section 9).

Definition A cadlag adapted process X on $[0, t_m]$ is said to be of class D if the family $\{X_u : u \in \mathfrak{T}[0, t_m]\}$ of random variables is equi-integrable.

Theorem (1) *Every cadlag submartingale X on $[0, t_m]$ such that the family $\{X_u : u \in \mathfrak{T}^f[0, t_m]\}$ is equi-integrable has a σ-additive Doléans measure.*

(2) *If for every $t \in [0, t_m]$ the σ-algebra $\mathfrak{F}_t$ contains all the P-null sets in $\mathfrak{F}_{t_m}$, every cadlag submartingale X with a σ-additive Doléans measure is of class D.*

Proof (1) In view of the σ-additivity lemma of Section 8.4, we have only to show that the Doléans function $d(X)$ of X fulfills assumptions (i), (ii), and (iii) of this lemma. As to (i), it is trivial since $d(X)$ is additive. Condition (ii) follows immediately from the fact that X is right continuous and from the equicontinuity property. For every increasing sequence of simple stopping times $u(n)$ we have

$$d(X)(]u(n), t_m]) = \int_{\{u(n) < t_m\}} (X_{t_m} - X_{u(n)}) dP,$$

and property (iii) is a straightforward consequence of the equi-integrability of the random variables $\{X_{t_m} - X_{u(n)} : n \in \mathbf{N}\}$.

(2) Let us define

$$u_n := \inf\{t : |X_t| > n\} \wedge t_m.$$

As already noticed, Doob's inequalities in Section 8.2 imply

$$\lim_{n\to\infty} P\{u_n < t_m\} = 0. \tag{8.5.1}$$

For every $v \in \mathfrak{T}[0, t_m]$, we set

$$H_v^n := \{|X_v| > n\}.$$

Since H_v^n belongs to $\mathfrak{F}_v$, $1_{H_v^n} 1_{]v, t_m]}$ is an adapted left continuous process and is therefore predictable. The set $]v, t_m] \cap (H_v^n \times]0, t_m])$ is predictable and can be written $]v, v_n]$ if we denote by v_n the stopping time $v + (t_m - v) 1_{H_v^n}$. From the proposition in Section 8.3(1), the σ-additivity of $d(X)$, and the theorem of Section 8.1 we derive

$$d(X)(]u, u']) = E(X_{u'} - X_u)$$

for every stochastic interval $]u, u'] \subset \Omega \times]0, t_m]$. Therefore

$$d(X)(]v, v_n]) = \int_{H_v^n} (X_{t_m} - X_v) dP,$$

and

$$\int_{H_v^n} |X_v| \, dP \leq \int_{H_v^n} |X_{t_m}| \, dP + d(X)(]v, v_n]). \tag{8.5.2}$$

But from the definition of u_n and v_n, it is clear that

$$]v, v_n] \subset]u_n, t_m] \qquad \text{and} \qquad H_v^n \subset \{u_n < t_m\},$$

and (8.5.2) then implies

$$\int_{\{|X_v|>n\}} |X_v|\, dP \leqslant \int_{\{u_n<t_m\}} |X_{t_m}|\, dP + d(X)(]u_n, t_m]). \tag{8.5.3}$$

The completeness assumption on the σ-algebras $\mathfrak{F}_t$ implies that every evanescent set A is contained in an element of $\mathcal{Q}$, which is negligible for $d(X)$. Relation (8.5.1) then implies

$$\lim_{n\to\infty} d(X)(]u_n, t_m]) = 0.$$

Therefore we obtain immediately from (8.5.3)

$$\lim_n \sup_{v\in\mathfrak{T}[0,t_m]} \int_{\{|X_v|>n\}} |X_v|\, dP = 0,$$

which expresses the equi-integrability of the random variables $\{X_v : v \in \mathfrak{T}[0, t_m]\}$.

Corollary *If for every $t \in [0, t_m]$ the σ-algebra $\mathfrak{F}_t$ contains all the P-null sets in $\mathfrak{F}_{t_m}$, the submartingale X is of class D if and only if the family $\{X_v : v \in \mathfrak{T}^f[0, t_m]\}$ of random variables is equi-integrable.*

8.6 Stopping Theorem for Martingales and Submartingales

Theorem *For every Banach-valued cadlag martingale M on $[0, t_m]$ and every stopping time u the stopped process $(M_{u\wedge t})_{t\in[0,t_m]}$ is a martingale. If X is a cadlag submartingale such that $\{X_u : u \in \mathfrak{T}^f[0, t_m]\}$ is equi-integrable (in particular, if X is positive; see the proposition in Section 8.3) for every stopping time u, the stopped process $(X_{u\wedge t})_{t\in[0,t_m]}$ is a submartingale. For every two stopping times u and v, $u \leqslant v$, one has $E(X_v - X_u \mid \mathfrak{F}_u) = 0$ (resp. $\geqslant 0$) if X is a martingale (resp. is a submartingale).*

Proof This theorem is a straightforward consequence of the general stopping theorem in Section 8.1, the equi-integrability of Section 8.3, and the existence of a σ-additive Doléans measure for M and X. The theorem of Section 8.1 implies indeed that the Doléans measure of $\overline{M}^u$ (the stopped process of M) is zero, while the Doléans measure of $\overline{X}^u$ is positive. The process $\overline{M}^u$ is therefore a martingale, while $\overline{X}^u$ is a submartingale. In the same way, for every $F \in \mathfrak{F}_u$, $E(1_F(X_v - X_u)) = 0$ (resp. $\geqslant 0$) if $v > u$ and X is a martingale (resp. a submartingale).

8.7 Equi-Integrability of Martingales

Theorem (1) *Let M be a Banach-valued martingale on $T = [0, t_\infty]$. Then M is of class D on $[0, t_\infty]$.*

(2) *Denoting by M^- the left continuous version of M, where M is a real*

martingale (see Section 1.17), *for every t we have*

$$E(M_{t_\infty} \mid \mathcal{F}_{t^-}) = M_t^- \text{ a.s.}$$

and the process M^- *is a martingale with respect to the σ-algebra* $\mathcal{F}_{t^-}$.

Proof (1) $\|M\|$ is a positive submartingale and we may apply the proposition of Section 8.3 and the corollary of Section 8.5.

(2) Let $0 < s < t$ and $F \in \mathcal{F}_{s-1/n}$. For every $k > n$ we have from the martingale property

$$\int_F M_{s-1/n}\, dP = \int_F M_{s-1/k}\, dP = \int_F M_{t-1/k}\, dP = \int_F M_{t_\infty}\, dP.$$

If we let k tend to $+\infty$, the equi-integrability property of the M_ts implies for every $F \in \bigcup_n \mathcal{F}_{s-1/n}$

$$\int_F M_{s^-}\, dP = \int_F M_{t^-}\, dP = \int_F M_{t_\infty}\, dP.$$

Since $\bigcup_n \mathcal{F}_{s-1/n}$ generates the σ-algebras $\mathcal{F}_{s^-}$, this proves part (2) of the theorem.

Notation From now on we shall write $E(M_{t_\infty} \mid \mathcal{F}_{t^-})$ for the *left continuous version of M*. In particular, if $F \in \mathcal{F}_{t_\infty}$, $E(1_F \mid \mathcal{F}_{t^-})$ will denote the left continuous version of $(E(1_F \mid \mathcal{F}_t))_{t \in T}$.

Remark When T is open, $T := [0, t_\infty[$, every real martingale M has an a.s. limit M_{t_∞} when $t \to t_\infty$. This follows from the basic argument in Sections 1.16 and 1.17. The equi-integrability property for M is then equivalent to the property

$$E(M_{t_\infty} \mid \mathcal{F}_t) = M_t \text{ a.s.}$$

When M is a Banach-valued martingale, the existence of the a.s. limit M_{t_∞}, when M is defined only on $[0, t_\infty[$, holds only when the Banach space has the Radon-Nikodym property. For a complete treatment of this point see [Met-6].

EXTENSIONS AND EXERCISES

1 Let X be a real cadlag supermartingale on $[0, t_m]$.

(1) If we set $Y_s = X_s - E(X_t \mid \mathcal{F}_s)$, (Y_s) is a positive supermartingale on $[0, t]$. For every increasing sequence (t_n) with $t_n \uparrow t$, the equality

$$E(X_{t^-}) = \lim_{n \to \infty} E(X_{t_n})$$

holds if and only if the random variables $\{X_{t_n}\}$ are equi-integrable.

(2) Use this to give examples of positive supermartingales (or negative submartingales) that are not equi-integrable.

(3) If $t \to E(X_t)$ is continuous on $[0, t_m]$ ($t_m < \infty$), the random variables $\{X_t : t \in [0, t_m]\}$ are equi-integrable (use (1) above and the proposition in Section 8.3(1)).

2 A submartingale may be equi-integrable without being of class D. An example, which is $1/\|B_t\|$, where B is a brownian motion in R^3, starting from a point different from 0, was given in [HeJ]. The proof uses special features of brownian motions (cf. also [Mey-1, Chap. IV]).

3 Here is an example which illustrates the fact that a submartingale, which is not of class D, has no σ-additive Doléans measure. Take $\Omega =]0, 1]$ and let P be the Lebesgue measure. Let t_n be a strictly increasing sequence $0 = t_0 < t_1 < \cdots < t_n < 1$. We call $\mathcal{F}_n^0$ the σ-algebra of subsets of Ω, generated by the intervals $]k2^{-n}, (k+1)2^{-n}]$, $0 \leqslant k < 2^n$. The σ-algebra $\mathcal{F}_t$ for $t \in [t_{n-1}, t_n[$ is the σ-algebra $\mathcal{F}_n^0$ completed by all the P-null sets. $\mathcal{F}_1$ is the Lebesgue σ-algebra of $]0, 1]$. We have

$$X_t = \begin{cases} 2^n & \text{on }]1 - 2^{-n}, 1] \quad \text{if } t \in [t_{n-1}, t_n[, \\ 0 & \text{on }]0, 1 - 2^{-n}] \quad \text{if } t \in [t_{n-1}, t_n[, \end{cases}$$

$$X_1 = 0 \qquad \text{everywhere.}$$

(1) X is a positive supermartingale, which is not of class D.

(2) Considering the sequence $]1 - 2^{-n}, 1] \times]t_n, 1]$ of sets in $\mathcal{A}$, check that the Doléans function $d(X)$ has no σ-additive extension to $\mathcal{P}$.

(3) If, instead of taking $\mathcal{F}_t$ complete, we had defined $\mathcal{F}_t = \mathcal{F}_n^0$ for $t \in [t_{n-1}, t_n[$ and $\mathcal{F}_1$ a Borel σ-algebra, would the conclusion of (2) remain true?

Comment This third question shows that if we drop the completeness hypothesis on the $\mathcal{F}_t$s, statement (2) of the theorem of Section 8.5 is no longer true. A study without completeness assumptions may be found in [Föl].

4 In Exercise 3 above X is a martingale on $[0, 1[$ but not on $[0, 1]$. Does there exist a random variable Y such that $Y_t = E(Y \mid \mathcal{F}_t)$ for $t \in [0, 1[$?

9 MEYER PROCESS AND DECOMPOSITION THEOREM

9.1 Generalities

In this section we suppose that $T = \mathbf{R}^+$, and we consider a complete probabilized stochastic basis $(\Omega, \mathcal{F}, P, (\mathcal{F}_t)_{t \in T})$. (We recall that for all t, $\mathcal{F}_t$ contains all the P-null sets in $\mathcal{F}$.) Moreover, we assume that $\mathcal{F}_t = \mathcal{F}_{t^+}$ (right continuity).

From the completeness assumption on the stochastic basis follows the property that every evanescent set is contained in an evanescent element of $\mathcal{A}$.

From Section 8.7 we recall that $(E(1_F \mid \mathfrak{F}_{t^-}))_{t\in T}$ is used to denote the *left continuous version* of the process $(E(1_F \mid \mathfrak{F}_t))_{t\in T}$.

DECOMPOSITION OF STOPPING TIMES

9.2 Predictable Stopping Times and Their Algebras $\mathfrak{F}_{u^-}$

The definition of predictable stopping times has been given in Section 1.14. We noticed that if u is predictable, the sets $]0, u[$ and $[u]$ are predictable.

If u is predictable, we call an increasing sequence $(u(n))_{n\in\mathbf{N}}$ of stopping times an "*announcing sequence for u*" if it converges to u and if $[u(n)](\omega) < u(\omega)$ for every ω such that $u(\omega) > 0$.

For a stopping time u the σ-algebra $\mathfrak{F}_{u^-}$ is defined as the σ-algebra generated by the following family of events:

$$\{F \cap \{t < u\} : F \in \mathfrak{F}_t, t \in \mathbf{R}^+\} \cup \mathfrak{F}_0.$$

We then have the following proposition.

Proposition (1) *For every stopping time u,* $\mathfrak{F}_{u^-} \subset \mathfrak{F}_u$.

(2) *If* $u(\omega) < v(\omega)$ *for all* ω *such that* $v(\omega) > 0$, *u and v being two stopping times, then* $\mathfrak{F}_u \subset \mathfrak{F}_{v^-}$.

(3) *For every increasing sequence u(n) of stopping times announcing a predictable stopping time u the σ-algebra* $\mathfrak{F}_{u^-}$ *is generated by* $\bigcup_n \mathfrak{F}_{u(n)}$.

(4) *If u is a predictable stopping time and if Z is an* $\mathfrak{F}_{u^-}$*-measurable random variable,* $Z 1_{[u,\infty[}$ *is a predictable process.*

Proof (1) Since for every $F \in \mathfrak{F}_t$, $s, t \in \mathbf{R}^+$, we have

$$F \cap \{t < u\} \cap \{u \leqslant s\} = F \cap \{t < u \leqslant s\} \in \mathfrak{F}_s,$$

the inclusion $\mathfrak{F}_{u^-} \subset \mathfrak{F}_u$ follows.

(2) Let F be an event in $\mathfrak{F}_u$. Since

$$F = F \cap \{v = 0\} \cup \left(\bigcup_{t\in Q} \{u \leqslant t < v\} \cap F \right),$$

we have only to prove that $F \in \mathfrak{F}_u$ implies

$$F \cap \{u \leqslant t < v\} \in \mathfrak{F}_{v^-}.$$

But $F \in \mathfrak{F}_u$ means $F \cap \{u \leqslant t\} \in \mathfrak{F}_t$. Therefore

$$F \cap \{u \leqslant t\} \cap \{t < v\} \in \mathfrak{F}_{v^-}$$

by definition of $\mathfrak{F}_{v^-}$.

(3) Following (1) and (2), it is sufficient to prove that $\mathcal{F}_{u^-}$ is generated by $\bigcup_n \mathcal{F}_{u(n)^-}$. This is immediate since

$$F \cap \{t < u\} = \bigcup_{n \in \mathbf{N}} F \cap \{t < u_n\}.$$

(4) If Z is a.s. $\mathcal{F}_{u^-}$-measurable, there exists a sequence $(Z_n)_{n>0}$ of random variables such that Z^n is $\mathcal{F}_{u(n)^-}$-measurable and Z_n converges a.s. to Z. Since the sequence $Z_n 1_{]u(n), \infty[}$ of predictable processes converges to $Z 1_{[u, \infty[}$ outside of an evanescent set, the process $Z 1_{[u, \infty[}$ is predictable.

9.3 Admissible Measures

A measure α on the σ-algebra $\mathcal{P}$ of predictable sets is said to be *admissible* if it is real (thus finite) and if $\alpha(B) = 0$ for every evanescent set B.

The Doléans measure of a process, when it exists (see Section 8 for conditions of existence) and is finite, is an admissible measure.

Let us remark that every right continuous increasing process A with $E(A_\infty - A_0) < \infty$ has a Doléans measure. This is an immediate consequence of the continuity properties of the integral which show that the formula

$$d_A(G) = E\left(\int_0^\infty 1_G(s, \cdot) dA_s(\cdot)\right)$$

defines a σ-additive measure on $\mathcal{P}$.

In this section to every admissible measure α we associate a (unique up to P-equivalence) predictable process A such that $\alpha = d_A$.

A fundamental step in this study is the "projection lemma" of Section 9.8.

9.4 Totally Inaccessible Stopping Times

Lemma and Definitions *Let u be a stopping time. The following two properties are equivalent:*

(i) $P\{w = u, u < \infty\} = 0$ *for every predictable stopping time w.*

(ii) *For every sequence $(v(n))_{n>0}$ of stopping times, increasing to a stopping time v, the sequence of the sets $(\{v(n) \geqslant u, u < \infty\})_{n>0}$ is, P-a.s., increasing to the set $\{v \geqslant u, u < \infty\}$.*

If these properties are satisfied, u is said to be totally inaccessible.

Proof We suppose first that condition (ii) is satisfied; let w be a predictable stopping time and $(v(n))_{n>0}$ be a sequence of stopping times that announces w; we have

$$\{v(n) \geqslant u, u < \infty\} \underset{n\to\infty}{\uparrow} \{v \geqslant u, u < \infty\}\ P\text{-a.s.}$$

and

$$\{v(n) \geqslant u, u < \infty\} \underset{n\to\infty}{\uparrow} \{w > u, u < \infty\}\ P\text{-a.s.}$$

Thus

$$P(\{w = u, u < \infty\}) = 0.$$

Now let $(v(n))_{n>0}$ be a sequence of stopping times that is increasing to v. For each integer n we put

$$v'(n) := v(n) \qquad \text{if} \quad v(n) < v$$

and

$$v'(n) := \infty \qquad \text{if} \quad v(n) = v.$$

It is easily seen that $v'(n)$ is a stopping time. The sequence $(v'(n))_{n>0}$ of stopping times is increasing to a stopping time v'. Since $v'(n) < v'$, as soon as $v'(n) < \infty$, v' is predictable and

$$\{v(n) \geqslant u, u < \infty\} \underset{n\to\infty}{\uparrow} (\{v \geqslant u\} \backslash \{v' = u\}).$$

If condition (i) is satisfied, we have

$$P(\{v' = u, u < \infty\}) = 0.$$

Thus condition (ii) holds.

9.5 Decomposition of a Stopping Time

Lemma *Let u be a stopping time. Then there exists a sequence $(v_n)_{n>0}$ of predictable stopping times and a totally inaccessible stopping time w such that*

(i) $[u] \subset \{[w] \cup (\bigcup_{n>0}[v_n])\}$.
(ii) $P\{v_j = v_k < \infty\} = 0$ *for every pair (j, k) of integers with $j \neq k$.*

Proof Let $\bar{b}$ be the supremum of the positive numbers b such that there exists a sequence of predictable stopping times v_n with

$$b = P\{\omega : \exists n > 0, u(\omega) = v_n(\omega)\}.$$

This supremum $\bar{b}$ is reached for a sequence $(v_n)_{n>0}$ of stopping times. Let w be the random variable defined by

$$w(\omega) := \begin{cases} u(\omega) & \text{if} \quad \forall n \quad u(\omega) \neq v_n(\omega), \\ \infty & \text{if} \quad \exists n \quad u(\omega) = v_n(\omega). \end{cases}$$

It is easily seen that w is a stopping time. The assumption

$$P\{w = v < \infty\} > 0$$

would contradict the definition of $\bar{b}$ and $(v_n)_{n>0}$. Therefore w is totally inaccessible. By substituting for v_n the following stopping times (if necessary),

$$v_n'(\omega) := \begin{cases} v_n(\omega) & \text{if} \quad \forall k < n \quad v_n(\omega) \neq v_k(\omega), \\ \infty & \text{if} \quad \exists k < n \quad v_n(\omega) = v_k(\omega), \end{cases}$$

we see that the sequence (v_n) can be chosen in such a way that (ii) holds. A stopping time u for which $[w]$ is evanescent is called *accessible*.

MEYER PROCESSES OF AN ADMISSIBLE MEASURE (REAL CASE)

9.6 Meyer Processes and the Class $\mathcal{C}$ of Processes

Definition (1) Let A be an increasing (real) process; A will be called a *Meyer process* if $A_0 = 0$, $E(A_\infty) < +\infty$, and A is a predictable right continuous process.

(2) We denote by $\mathcal{C}$ the set of the processes A which satisfy the following properties:

(i) A is real, adapted, increasing, and right continuous;
(ii) $A_0 = 0$ and $E(A_\infty) < +\infty$;
(iii) for every element F of $\mathfrak{F}$ and for every stopping time u we have

$$E(1_F A_u) = \int_{]0,\,u]} E(1_F \mid \mathfrak{F}_{t^-})\, d\alpha,$$

where α is the Doléans measure associated with the process A.

If A is an element of $\mathcal{C}$ and u is a stopping time, let us call $\bar{A}^u$ the process A stopped at the stopping time u (cf. Section 1.12); it is also an element of $\mathcal{C}$ because the Doléans measure α^u associated with $\bar{A}^u$ is defined by $\alpha^u(B) = \alpha(]0, u] \cap B)$ for every predictable set B.

We clearly have the same property if A is a Meyer process: For every stopping time u, $\bar{A}^u$ is a Meyer process.

We shall see in Section 9.12 that definitions (1) and (2) are equivalent: $\mathcal{C}$ is the class of Meyer processes.

9.7 Construction of A

Theorem *Let α be an admissible measure. Then there exists an element A of $\mathcal{C}$ such that α is the Doléans measure of A; this process A is unique up to P-equivalence.*

Proof (1) The uniqueness of A up to a modification follows from the condition in Section 9.6(iii); but since A is right continuous, it is determined up to P-equivalence.

(2) Let t be an element of T; for every element H in $\mathcal{F}_t$ we put

$$v_t(H) := \int_{]0,t]} E(1_H \mid \mathcal{F}_{s^-})d\alpha.$$

The function $v_t(\cdot)$ is σ-additive (Lebesgue theorem) and dominated by P; then we can set $V_t := dv_t/dP$, and this defines up to modification an increasing process V which is moreover right continuous in the mean. Let us take for A a right continuous process which is a modification of V: Such a process exists from the above property of V.

(3) Let Y be an element of $L_\infty(\Omega, \mathcal{F}, P)$; we may write

$$\int_{]0,t]} E(Y \mid \mathcal{F}_{s^-})d\alpha = \int_\Omega Y(w)v_t(d\omega).$$

Indeed, if $Y = 1_F$ with $F \in \mathcal{F}$, this equality follows from the definition of v, and in the general case it is obtained from a classical linearity and density argument, and therefore $E(A_t 1_H) = E(A_t E(1_H \mid \mathcal{F}_t))$ For all $H \in \mathcal{F}$.

(4) A is an adapted process since we have

$$v_t(H) = \int_{]0,t]} E(1_H \mid \mathcal{F}_{s^-})d\alpha = \int_{]0,t]} E[E(1_H \mid \mathcal{F}_t) \mid \mathcal{F}_{s^-}]d\alpha$$
$$= \int_\Omega E(1_H \mid \mathcal{F}_t)v_t(d\omega)$$

(cf. point (3) above).

(5) Let $u = \sum_{i \in I} t(i)1_{G(i)}$ be a simple stopping time (this means for disjoint $G(i)$s that if $i \in I$, $G(i) \in \mathcal{F}_{t(i)}$). We then have

$$E[1_F(A_\infty - A_u)] = \sum_{i \in I} (v_\infty - v_{t(i)})(F \cap G(i))$$
$$= \sum_{i \in I} \int_{]t(i),\infty]} E[1_F 1_{G(i)} \mid \mathcal{F}_{s^-}]d\alpha$$
$$= \sum_{i \in I} \int_{]t(i),\infty]} 1_{G(i)} E(1_F \mid \mathcal{F}_{s^-})d\alpha$$
$$= \int_{]u,\infty]} E(1_F \mid \mathcal{F}_{s^-})d\alpha.$$

Then the property of Section 9.6(iii) is satisfied if u is a simple stopping time; thus we have this same property if u is a general stopping time by considering such a stopping time as the limit of a decreasing sequence of simple stopping times (see Section 1.11).

9.8 "Projection" Lemma

Lemma *Let u be a totally inaccessible stopping time. Let B (resp. C) be the process defined by $B := 1_{[u,\infty[}$ (resp. $C := 1_{]u:\infty[}$). For each integer n let B^n (resp. C^n) be the right (resp. left) continuous process defined by*

$$B_t^n = E(B_{(k+1)2^{-n}} \mid \mathcal{F}_{t^+}) \qquad \text{if} \quad k2^{-n} \leqslant t < (k+1)2^{-n},$$

$$C_t^n = E(B_{(k+1)2^{-n}} \mid \mathcal{F}_{t^-}) \qquad \text{if} \quad k2^{-n} < t \leqslant (k+1)2^{-n}.$$

When n goes to infinity, the sequence $(B^n)_{n>0}$ (resp. $(C^n)_{n>0}$) converges P-a.s. uniformly to the process B (resp. C).

Proof We should first remark that this lemma is a corollary of the properties of "predictable projections" as studied in [Del]; we give a direct proof of it since it will meet our needs here and will enable us to dispense with "section and projection theorems" and the "capacitability theorem" as done in [Del].

(1) The processes B^n and C^n are defined up to P-equivalence, the sequences $(B^n)_{n>0}$ and $(C^n)_{n>0}$ are decreasing, and for every integer n, $B^n \geqslant B$. We take $B_\infty^n = B_\infty = 0$. Let ϵ be a positive number; for each integer n let $v(n)$ be the stopping time defined by

$$v(n) := \inf\{t : B_t^n - B_t > \epsilon\}.$$

The sequence $(v(n))_{n>0}$ is increasing to a stopping time v.

(2) For convenience of notation we put

$$\bar{u} := 2^n u, \qquad \bar{v} := 2^n v, \qquad \overline{v(n)} := 2^n v(n),$$

$$D := \{\omega : k \leqslant \overline{v(n)}(\omega)\},$$

k being temporarily fixed, and

$$a(n,k) := E(B_{v(n)}^n 1_{k \leqslant \overline{v(n)} < k+1}).$$

When $k2^{-n} \leqslant t$, $1_D B^n$ is a submartingale (because $D \in \mathcal{F}_{k2^{-n}}$), then (cf. the stopping theorem of Section 8.6) we have

$$\begin{aligned} a(n,k) &\leqslant E(B_{(k+1)2^{-n}} 1_D) \\ &\leqslant E(1_{\{\bar{u} \leqslant k+1\}} 1_D) \end{aligned}$$

But $v(n) < \infty$ implies $v(n) < u$ and

$$\begin{aligned} a(n,k) &\leqslant E\left[1_{\{\bar{u} \leqslant k+1\}} 1_D 1_{\{\overline{v(n)} < k+1\}}\right] \\ &\leqslant E\left[1_{\{\bar{u} \leqslant \bar{v}(n)+1\}} 1_{\{k \leqslant \bar{v}(n) < k+1\}}\right]. \end{aligned}$$

Thus we have

$$E[B^n_{v(n)}] = \sum_{k\geq 0} E[a(n,k)]$$
$$\leq E[1_{\{u<\infty\}} 1_{\{\bar{u}\leq \overline{v(n)}+1\}}]$$
$$\leq P(\{\bar{u} \leq \bar{v}+1, u<\infty\}).$$

Then

$$\lim_{n\to\infty} E[B^n_{v(n)}] \leq P(\{u \leq v, u<\infty\}).$$

(3) The definition of $v(n)$ implies

$$E[B^n_{v(n)} - B_{v(n)}] \geq \epsilon P(\{v(n)<\infty\}) \geq \epsilon P(\{v<\infty\}).$$

If we consider the limit of this inequality when n goes to infinity, we obtain (cf. (2) above)

$$P(\{u \leq v, u<\infty\}) - \lim P(\{u \leq v(n), u<\infty\}) \geq \epsilon P(\{v<\infty\}).$$

But

$$\{u \leq v(n), u<\infty\}\uparrow\{u \leq v, u<\infty\}, \qquad \text{as} \quad n\uparrow\infty \quad \text{(cf. Section 9.4);}$$

thus

$$\epsilon(P\{v<\infty\}) = 0.$$

This last equality proves that the sequence $(B^n)_{n>0}$ converges P-a.s. uniformly for each sample function to the process B. But for every integer n, C^n is the left continuous process associated with the right continuous process B^n; thus the sequence $(C^n)_{n>0}$ converges P-a.s. uniformly for each sample function to the process C defined by $C_t = B_{t^-}$, i.e., $C = 1_{]u,\infty[}$.

9.9 Case in Which A Is Continuous

Proposition *Let A be an element of $\mathcal{C}$ and α be its Doléans measure. We suppose that for every predictable stopping time u, $\alpha([u]) = 0$. Then A is* a.s. *continuous and therefore a Meyer process.*

Proof (1) Let u be a predictable stopping time and $(u(n))_{n>0}$ be a sequence announcing u. We have

$$0 = \alpha([u]) = \lim_{n\to\infty} \alpha(]u(n),u]) = \lim_{n\to\infty} E[A_u - A_{u(n)}]$$
$$= E(A_u - A_{u^-}).$$

(2) Let u be a totally inaccessible stopping time. We define the sequence $(C^n)_{n>0}$ of processes as in Section 9.8. For each pair (n,k) of integers we put

$$D(n,k) = \{k2^{-n} < u \leq (k+1)2^{-n}\}$$

and

$$w(n) = \sum_{k \geq 0} (k+1)2^{-n} 1_{D(n,k)}.$$

We have

$$E(A_u - A_{u^-}) = \lim_{n\to\infty} \left\{ \sum_{k \geq 0} E\left[1_{D(n,k)}(A_{(k+1)2^{-n}} - A_{k2^{-n}})\right]\right\}.$$

This equality and the property in Section 9.6(iii) imply

$$E(A_u - A_{u^-}) = \lim_{n\to\infty} \int \left[C^n - 1_{]w(n),\infty[}\right] d\alpha,$$

and this limit is equal to zero (cf. Section 9.8).

(3) Then for every stopping time u we have

$$E(A_u - A_{u^-}) = 0$$

(cf. Section 9.5 and (1)–(2) above). For each $\epsilon > 0$ let us consider the stopping time u defined by

$$u := \inf\{t : (A_t - A_{t^-}) > \epsilon\}.$$

It is clear that $u = +\infty$ from what precedes. Therefore A is continuous.

9.10 Case in Which $\alpha([u]) = \alpha(\Omega')$

Proposition *Let A be an element of $\mathcal{C}$. Let u be a predictable stopping time. We suppose that $\alpha([u]) = \alpha(\Omega')$. Then the process A is predictable (i.e., A is a Meyer process) and has the form $A = A_\infty 1_{[u,\infty[}$.*

Proof We have

$$E(A_\infty - A_0) = \alpha(\Omega') = \alpha([u]) = E(A_u - A_{u^-})$$

(cf. Section 9.9(1)). This means that A has one jump on the stopping time u and is constant elsewhere. Let $(u(n))_{n>0}$ be an announcing sequence for u. Let F be an element of $\mathcal{F}$. We have (cf. Section 9.6(iii))

$$E(1_F A_u) = \int E(1_F \mid \mathcal{F}_{t^-}) 1_{]0,u]}\, d\alpha.$$

But the martingale $E(1_F \mid \mathcal{F}_t)$ stopped at $u(n)$ is P-equivalent to the martingale $E[E(1_F \mid \mathcal{F}_{u^-}) \mid \mathcal{F}_t]$ stopped at $u(n)$; thus we have the same property when we stop these martingales at u. From this we derive

$$E(1_F A_u) = \int E\left[E(1_F \mid \mathcal{F}_{u^-}) \mid \mathcal{F}_{t^-}\right] 1_{]0,u]}\, d\alpha$$
$$= E\left[E(1_F \mid \mathcal{F}_{u^-}) A_u\right].$$

This proves that the random variable A_u is $\mathcal{F}_{u^-}$-measurable and $A_u 1_{[u,\infty[}$ is predictable (cf. the end of Section 9.2).

9.11 Integration of a Martingale with Respect to an Increasing Process

Proposition *Let M be a uniformly bounded right continuous martingale and A be an adapted integrable increasing right continuous process. For every element t of T we have*

$$E\left[\int_{]0,\,t]} M_s\,dA_s\right] = E[M_t(A_t - A_0)].$$

Proof We note $T_t := (T \cap [0, t])$. Let $(T_n)_{n>0}$ be an increasing sequence of finite subsets of T_t such that $\bigcup_{n>0} T_n$ is dense in T_t, $t \in T_1$, and $0 \in T_1$. For each integer n let $(t^n(k))_{1 \leqslant k \leqslant q}$ be the increasing family of elements of T_n and let M^n be the process defined by

$$M^n = \sum_{k=1}^{q-1} M_{t^n(k+1)} 1_{]t^n(k),\, t^n(k+1)]}.$$

The sequence of processes $(M^n)_{n>0}$ converges to the process M; by the dominated convergence theorem it is sufficient to prove the equality for each process M^n. But we have from the definition of M^n

$$\begin{aligned} E\left[\int_{]0,\,t]} M_s^n\,dA_s\right] &= \sum_{k=1}^{q-1} E\{M_{t^n(k+1)}[A_{t^n(k+1)} - A_{t^n(k)}]\} \\ &= \sum_{k=1}^{q-1} E\{M_t[A_{t^n(k+1)} - A_{t^n(k)}]\} \\ &= E[M_t(A_t - A_0)]. \end{aligned}$$

9.12 Meyer Process: Existence and Uniqueness (Real Case)

Theorem (a) *If A and B are two increasing Meyer processes which have the same Doléans measure (i.e., $(A - B)$ is a martingale), then A and B are P-equivalent.*

(b) *If α is a positive admissible measure, there exists a Meyer process A such that α is the Doléans measure of the process A. Moreover, this process A is continuous (up to P-equivalence) if and only if $\alpha([u]) = 0$ for all predictable stopping times u.*

(c) *An increasing process A is a Meyer process if and only if A is an element of $\mathcal{C}$.*

Proof (a) Let A and B be two Meyer processes such that $(A - B)$ is a martingale M and let us suppose first that A and B are uniformly bounded.

We have (cf. Section 9.11)

$$E(M_t A_t) = E\left(\int_{]0,\,t]} M_s\, dA_s\right).$$

If α is the common Doléans measure of A and B, and M is a predictable process, we may write

$$E\left(\int_{]0,\,t]} M_s\, dA_s\right) = \int_{]0,\,t]} M_s\, d\alpha = E\left(\int_{]0,\,t]} M_s\, dB_s\right).$$

By applying the proposition in Section 9.11 we obtain

$$0 = E\left[M_t(A_t - B_t)\right] = E\left(M_t^2\right),$$

which proves that M is evanescent. Now we suppose that A and B are two Meyer processes such that $(A - B)$ is a martingale, without the assumption that A and B are uniformly bounded. For every integer n we consider the predictable set $C(n)$ where $(A_t + B_t) \geqslant n$ and $D(n) := \Omega' \setminus C(n)$; we put

$$A^n := 2n 1_{C(n)} + A 1_{D(n)}$$

and

$$B^n := 2n 1_{C(n)} + B 1_{D(n)}.$$

The processes A^n and B^n are two bounded Meyer processes which have the same Doléans measures; thus A^n and B^n are P-equivalent. This proves that A and B are also P-equivalent.

(b) Let α be a positive admissible measure. To prove that there exists a Meyer process A such that α is its Doléans measure, we first prove that $\alpha = \mu + \sum_{n>0} \beta_n$, where μ is as in Section 9.9 and β_n is as in Section 9.10. We consider the supremum $\bar{c}$ of the positive numbers $\bar{b}$ such that

[i] there exists a sequence $(u(n))_{n>0}$ of predictable stopping times with disjoint graphs and a sequence $(\beta_n)_{n>0}$ of Doléans measure such that

$$\bar{b} = \sum_{n>0} \beta_n(\Omega'), \qquad \sum_{n>0} \beta_n(\cdot) \leqslant \alpha(\cdot),$$

and for every integer n,

$$\beta_n(\Omega') = \beta_n([u(n)]).$$

It is easily seen that this supremum $\bar{c}$ is reached for a sequence $(u(n), \beta_n)_{n>0}$ satisfying condition [i], with $\bar{b} = \bar{c}$. We put $\mu := \alpha - \sum_{n>0} \beta_n$. Let D and B^n be the processes belonging to $\mathcal{C}$ associated with μ and β_n, respectively, as built in Section 9.7. These processes are also Meyer processes (cf. Sections 9.9 and 9.10); we have

$$\sum_{n>0} E(B^n_\infty) = \sum_{n>0} \beta_n(\Omega') < +\infty.$$

Moreover, we define the Meyer processes $A^n := D + \sum_{k=1}^n B^k$. Since

$$\sup_t (A_t^{n+k} - A_t^n) \leqslant \sum_{r=n+1}^{n+k} B_\infty^r,$$

by using the Borel–Cantelli lemma, we may extract from A^n a subsequence which converges P-a.s. uniformly for every sample function to a process A, which, at the same time, is a Meyer process and belongs to $\mathcal{C}$. Moreover, the Doléans measure of A is equal to $\mu + \sum_{n>0} \beta_n = \alpha$. If A is continuous, for every predictable stopping time u, $\alpha([u]) = E(A_u - A_{u^-}) = 0$. Conversely, if $\alpha([u]) = 0$ for every predictable stopping time, A is continuous, as was seen in Section 9.9.

(c) Let A' be an increasing right continuous process with $A'_0 = 0$ and $E(A'_\infty) < +\infty$. Let α be its Doléans measure. Let A be the associated process as built above; A is a Meyer process and belongs to $\mathcal{C}$. Then A is P-equivalent to A' if A' is a Meyer process or if A' belongs to $\mathcal{C}$, and this proves statement (c).

MEYER PROCESS: VECTOR CASE

9.13 Banach Space Valued Meyer Processes

Definition Let A be a Banach-valued process; we say that A is a Meyer process if A is a predictable cadlag process with bounded integrable variation for each sample path and is such that $A_0 = 0$.

9.14 Existence and Uniqueness

Proposition *Let* $\mathbf{B}$ *be a Banach space which is reflexive or, more generally, which satisfies the "Radon–Nikodym property".*[2] *Let* α *be a* $\mathbf{B}$*-valued* σ*-additive measure defined on the* σ*-algebra of predictable sets. We assume that the total variation* $|\alpha|$ *of* α *is finite and is an admissible measure (see Section 9.3). We denote by* Q *the Radon-Nikodym density of* α *with respect to* $|\alpha|$ *and by* $\bar{A}$ *the increasing Meyer process associated with* $|\alpha|$*. Let* A *be the* $\mathbf{B}$*-valued predictable process defined by*

$$A_t = \int_{]0,\,t]} Q_s \, d\bar{A}_s,$$

where this integral is a classical Stieltjes integral. Then A *is called the Meyer process associated with* α*. It is the unique Meyer process up to* P*-equivalence such that* α *is the Doléans measure of* A*.*

[2] A Banach space $\mathbf{B}$ is said to possess the Radon–Nikodym property if for every measure space $(\Omega, \mathfrak{F}, \mu)$ and any $\mathbf{B}$-valued measure ν on $(\Omega, \mathfrak{F})$ with finite variation, such that $\mu(A) = 0 \Rightarrow \nu(A) = 0$, there exists a $\mathbf{B}$-valued density of ν with respect to μ. Every reflexive separable Banach space has the Radon–Nikodym property.

Proof Q is well defined by the assumption on **B**. It is very easy to see that A is a Meyer process and that α is the Doléans measure of A. The uniqueness of $\langle A, x\rangle$ for every element x of the dual **B**$'$ of **B** (see Section 9.12) implies the uniqueness of A.

DECOMPOSITION OF LOCAL MARTINGALES

9.15 A Locally Square Integrable Process

Lemma *Let **H** be a Hilbert space and Z an **H**-valued cadlag adapted π-process uniformly bounded by d (i.e., $\|Z_t(\omega)\| \leqslant d$ for every $(\omega, t) \in (\Omega \times T)$). We assume that Z admits a Doléans measure α, the total variation $|\alpha|$ of which is an admissible measure. We denote by A and $\bar{A}$ the Meyer processes associated with α and $|\alpha|$, respectively. Then A, $\bar{A}$, and $(A - Z)$ are locally square integrable (i.e., there exists an increasing sequence $(u(n))_{n>0}$ of stopping times such that $\lim_n u(n) = +\infty$ and for every n and every t*

$$E\left[\|A_{u_n \wedge t}\|^2\right] < \infty, \qquad E\left(\bar{A}^2_{u_n \wedge t}\right) < \infty, \qquad E\left[\|(A - Z)_{u_n \wedge t}\|^2\right] < \infty.$$

Proof Since $\|A\| \leqslant \bar{A}$, it is sufficient to prove that $\bar{A}$ is locally square integrable.

(1) Let b be a positive number and u the stopping time defined by

$$u := \inf\left\{t : \bar{A}_t > b\right\}.$$

We assume that

$$P\left(\left\{\omega : \exists s < t, \bar{A}_s = \bar{A}_t = b\right\}\right) = 0.$$

In this case

$$[u] = \{(\omega, t) \colon \bar{A}_t(\omega) \geqslant b\} \backslash]u, \infty],$$

and $[u]$ is a predictable set. We have (because clearly $\bar{A}_u - \bar{A}_{u^-} = \|A_u - A_{u^-}\|$ a.s.)

$$x := E\left[\left(\bar{A}_u - \bar{A}_{u^-}\right)^2\right] = E\left[(A_u - A_{u^-})^2\right]$$
$$= E\left\{\int 1_{[u]}\langle A_s - A_{s^-}, dA_s\rangle\right\},$$

where $\langle \cdot, \cdot \rangle$ denotes the scalar product in **H**. Now, $(1_{[u]}(A_s - A_{s^-}))_{s\in T}$ is a predictable process and A and Z have the same Doléans measure; thus we have

$$x = E\left\{\int 1_{[u]}\langle A_s - A_{s^-}, dZ_s\rangle\right\} \leqslant 2dE(\|A_u - A_{u^-}\|) < +\infty$$

and

$$E(\|A_u\|^2) < +\infty.$$

(2) Let $(b(n))_{n>0}$ be an increasing sequence of positive numbers such that

$$\lim_{n\to\infty} b(n) = +\infty$$

and for every integer n

$$P\left(\left\{\omega : \exists s < t, \bar{A}_s = \bar{A}_t = b(n)\right\}\right) = 0.$$

Such a sequence exists according to the denumerability of the set D:

$$D := \left\{d : d \geqslant 1 \text{ and } P\left(\left\{\omega : \exists s < t, \bar{A}_s(\omega) = \bar{A}_t(\omega) = d\right\}\right) > 0\right\}.$$

For every integer n we define

$$u(n) = \inf\left\{t : \bar{A}_t > b(n)\right\}.$$

According to (1),

$$E\left(\bar{A}^2_{u(n)}\right) < +\infty$$

and $\bar{A}$ is locally square integrable.

9.16 Decomposition of a Local Martingale

Definition A process M will be called a *local martingale* if it is locally a martingale. More precisely, there exists an increasing sequence $u(n)$ of stopping times such that $\lim_n u(n) = +\infty$ and for every n the stopped process $\overline{M}^{u(n)}$ (see Section 1.12) is a martingale. From the stopping theorem of Section 8.6, every martingale is trivially a local martingale.

Theorem *Let* $\mathbf{H}$ *be a Hilbert space and* M *an* $\mathbf{H}$*-valued local martingale. Then there exists an* $\mathbf{H}$*-valued cadlag locally square integrable martingale* W *and an* $\mathbf{H}$*-valued cadlag adapted process* V *with finite variation on every bounded interval* $[0, t] \subset T$ *such that* $M = W + V$.

Proof Let $(u(n))_{n>0}$ be a sequence of stopping times such that for every integer n the process M stopped at $u(n)$ is a martingale. Let $(v(n))_{n>0}$ be the sequence of stopping times defined by

$$v(0) := 0$$

and

$$v(n) := u(n) \wedge \inf\{t : |M_t| > n\}.$$

We define recurrently

$$M^0 := M_0$$

and

$$M^n := (M - M^{n-1})1_{]v(n-1), v(n)]} + \left(M_{v(n)} - M^{n-1}_{v(n)}\right)1_{]v(n), \infty[},$$

$$B^n := \left[M^n_{v(n)} - M^n_{v(n)^-}\right]1_{[v(n), \infty[}$$

and

$$Z^n := B^n - M^n.$$

Let α^n be the Doléans measure of the process B^n and A^n be the Meyer process associated with this Doléans measure (the total variation of α^n is clearly an admissible measure). Now α^n is the Doléans measure of the uniformly bounded process Z^n, and $(A^n - Z^n)$ is locally square integrable (see Section 9.15). Then we have

$$M - M_0 = \sum_{n>0} M^n = \sum_{n>0} (B^n - A^n) + \sum_{n>0} (A^n - Z^n)$$
$$= V - W$$

where

$$V := \sum_{n>0} (B^n - A^n)$$

is a process of finite variation and

$$W := \sum_{n>0} (A^n - Z^n)$$

is a locally square integrable martingale (for every integer n the process W stopped at $v(n)$ is a finite sum $\sum_{k\leqslant n}(A^k - Z^k)$ of locally square integrable martingales).

9.17 Case in Which $\mathcal{F}_t \neq \mathcal{F}_{t^+}$

Remark We supposed the family $(\mathcal{F}_t)_{t\in T}$ to be right continuous. In the general case it is always possible to consider the family $(\mathcal{F}_{t^+})_{t\in T}$ and to use the results of Section 9; in this case indeed a Meyer process with respect to the family $(\mathcal{F}_{t^+})_{t\in T}$ is also a Meyer process with respect to the family $(\mathcal{F}_t)_{t\in T}$ (cf. Section 1.10).

EXTENSIONS AND EXAMPLES

The details of what follows are left as exercises for the reader.

1 There is a converse to statement (4) in the proposition in Section 9.2. Let us as usual call X^- the process $X^-(t) = \lim_{s\uparrow t,\, s<t} X(s)$ and $\Delta X = X - X^-$ when X is a cadlag process. Let u be a T-valued stopping time. It is very easy to show from the definition of $\mathcal{F}_{u^-}$ that X_u^- is $\mathcal{F}_{u^-}$-measurable for every adapted cadlag process X. *If X is predictable, X_u itself is $\mathcal{F}_{u^-}$-measurable*, and ΔX_u is therefore $\mathcal{F}_{u^-}$-measurable too (the property is true for $X = 1_{]u,v]}$: use then a classical measure theoretic extension argument).
2 Let N be a Poisson process adapted to the given stochastic basis (see Section 1.5). The Doléans measure of N being $P \otimes l$, where l is the Lebesgue measure on T, the continuous increasing process $\tilde{N}_t = t$ (nonran-

dom), which generates the same Doléans measure as $\tilde{N}$, is clearly the Meyer process of this measure. The continuity of N shows (see Section 9.12) that *the jumps of N occur at totally inaccessible stopping times.*

3 Let u be a T-valued stopping time ($T := [0, t_\infty]$). If h is an $\mathfrak{F}_u$-measurable **H**-valued random variable with $E\|h\| < \infty$, the Doléans measure of $X := h1_{[u, t_\infty]}$ exists. We call $\tilde{X}$ the associated Meyer process. Show that *if u is predictable*, $\tilde{X} := E(h \mid \mathfrak{F}_{u^-})1_{[u, t_\infty]}$. (Note that $\tilde{X}$ is in fact predictable and that the σ-algebra $\mathfrak{P} \cap [u]$ consists of the sets $\{H \times [0, t_\infty] \cap [u] : H \in \mathfrak{F}_{u^-}\}$. One may also note that $h1_{[u, t_\infty]}$ is a martingale if and only if $E(h \mid \mathfrak{F}_{u^-}) = 0$).

4 (Cf. [Del]). We put $\Omega := [0, 1]$, $\mathfrak{F} :=$ σ-algebra of the Borel sets of Ω, P: probability on $(\Omega, \mathfrak{F})$, $T := [0, 1]$ for every element t of T, $\mathfrak{B}_t :=$ σ-algebra of all the Borel sets contained in $[0, t]$, $\mathfrak{F}_t :=$ σ-algebra on Ω generated by $\mathfrak{B}_t$; more precisely, a subset A of Ω belongs to $\mathfrak{F}_t$ if and only if A belongs to $\mathfrak{B}_t$ or if $A = B \cup]t, 1]$, where B belongs to $\mathfrak{B}_t$. Then $(\Omega, \mathfrak{F}, P, (\mathfrak{F}_t)_{t \in T})$ is a stochastic basis.

(1) Is the family $(\mathfrak{F}_t)_{t \in T}$ right continuous?

(2) Let v be a stopping time; prove that $v(s) < s$ and $v(t) < t$ imply that $v(s) = v(t)$ (almost surely).

(3) Let u be the T-valued random variable defined on Ω by $u(s) := s$. Is u a stopping time? Is u a predictable stopping time? Is u a totally inaccessible stopping time?

5 We consider $T := [0, 1]$, $\Omega := [0, 1]$, $\mathfrak{F} :=$ σ-algebra of the Borel sets of Ω, $\mathfrak{F}_t := \{\emptyset, \Omega\}$ if $t < \frac{1}{2}$ and $\mathfrak{F}_t := \mathfrak{F}$ if $t \geqslant \frac{1}{2}$, M_1 an element of $L_1(\Omega, \mathfrak{F}, \mathfrak{P})$ such that $E(M_1) = 0$, $M_t = M_1$ if $t \geqslant \frac{1}{2}$ and $M_t = 0$ if $t < \frac{1}{2}$. We suppose $E(|M_1|^2) = +\infty$. Is $(M_t)_{t \in T}$ a locally square integrable martingale? Is $[M, M]$ a locally integrable process? Is it possible to define a Meyer process associated with $[M, M]$?

10 π^*-PROCESSES AND SEMIMARTINGALES

In this section, we prove that every semimartingale on $T := [0, t_\infty]$ is a π^*-process. First, we need more information about the structure of square integrable martingales.

SQUARE INTEGRABLE MARTINGALES

10.1 The L^2-Space of Square Integrable Martingales

We consider processes which are indexed by $T := [0, t_\infty]$ with t_∞ finite or not.

We shall denote by $\mathfrak{M}_T^2(\mathbf{H})$ the space of cadlag square integrable martingales on T, i.e., all the martingales M on T such that $E(\|M_{t_\infty}\|^2)$

$< \infty$. As we know from Section 8.7, an element of $\mathfrak{M}^2_T(\mathbf{H})$ is uniquely defined up to P-equivalence by the random variable M_{t_∞}. It is therefore immediately clear that the bilinear form

$$(M, N) \rightsquigarrow E(M_{t_\infty} N_{t_\infty})^3$$

is a scalar product on $\mathfrak{M}^2_T(\mathbf{H})$, turning $\mathfrak{M}^2_T(\mathbf{H})$ into a complete Hilbert space, if we agree to identify P-equivalent processes.

Two elements M and N of $\mathfrak{M}^2_T(\mathbf{H})$ are therefore orthogonal in this Hilbert space if $E(M_{t_\infty} N_{t_\infty}) = 0$.

This property is clearly weaker than the property expressed by $E(M_u N_u) = 0$ for all T-valued stopping times u, which is equivalent to the nullity of the Doléans measure of MN when $(M_0 N_0) = 0$ a.s. It is usually said that *M and N are strongly orthogonal when $E(M_u N_u) = 0$ for all T-valued stopping times u.*

Since $M_t N_t = \mathrm{Tr}(M_t \otimes N_t)$, $E(M_u \otimes N_u) = 0$ implies $E(M_u N_u) = 0$.

Examples (1) Let u be a stopping time and $h \in L^2_{\mathbf{H}}(\Omega, \mathfrak{F}_u, P)$. We consider the process $A := h 1_{[u, t_\infty]}$. It is a process with finite Doléans measure d_A, and we call $\tilde{A}$ the associated Meyer process. The process $N := A - \tilde{A}$ is then a martingale, and we show that $N \in \mathfrak{M}^2_T(\mathbf{H})$. Let us first assume that u is predictable. Then (see Section 9.18(3))

$$\tilde{A} = E(h \mid \mathfrak{F}_{u^-}) 1_{[u, t_\infty]},$$

and therefore

$$E\|N_{t_\infty}\|^2 \leqslant 4E\|h\|^2. \tag{10.1.1}$$

If u is totally inaccessible, $\tilde{A}$ is continuous (see Section 9.9). A classical formula for Stieltjes integrals gives

$$\begin{aligned} E(\|\tilde{A}_{t_\infty}\|^2) - 2E\left(\int_{[0, t_\infty]} \tilde{A}_s \, d\tilde{A}_s\right) &\\ = 2E\left(\int_{[0, t_\infty]} \tilde{A}_s \, dA_s\right) & \qquad \text{(from the definition of } \tilde{A}) \\ = 2E(\tilde{A}_u h). & \end{aligned}$$

Noticing that $d_A(]u, t_\infty]) = 0$, we obtain $\tilde{A}_u = \tilde{A}_{t_\infty}$. Then using the Schwarz inequality, we get

$$E(\|\tilde{A}_{t_\infty}\|^2) \leqslant 2\sqrt{E\|\tilde{A}_{t_\infty}\|^2} \sqrt{E\|h\|^2},$$

[3] To avoid confusion with the Meyer process denoted by $\langle\, , \rangle$ (see Section 10.2), we shall write simply xy for the scalar product of x and y, returning to the notation $\langle x, y\rangle_{\mathbf{H}}$ when the separable Hilbert space $\mathbf{H}$ has to be specified.

and therefore

$$E\|\tilde{A}_{t_\infty}\|^2 \leqslant 4E\|h\|^2. \tag{10.1.2}$$

(2) Let us consider now the process $A := h1_{[u, t_\infty]}$, $h \in L^2_{\mathbf{H}}(\Omega, \mathcal{F}_u, P)$ when u is a general stopping time. We take the decomposition

$$[u] = [u_0] \cup [v],$$

where u_0 is totally inaccessible and $[v] \subset \bigcup_n [v_n]$, the v_n being predictable with disjoint graphs. We have

$$A = \Delta A_{u_0} 1_{[u_0, t_\infty]} + \sum_n \Delta A_{u_n} 1_{[u_n, t_\infty]}$$

with

$$\|h\|^2 = \|h\|^2 1_{\{u = u_0\}} + \sum_n \|\Delta A_{u_n}\|^2 \tag{10.1.3}$$

since

$$\tilde{A} = \tilde{A}_0 + \sum_n E(\Delta A_{u_n} \mid \mathcal{F}_{u_n^-}) 1_{[u_n, t_\infty]},$$

where $\tilde{A}_0$ denotes the Meyer process of $\Delta A_{u_0} 1_{[u_0, t_\infty]}$. From (10.1.1)–(10.1.3), we deduce

$$E\|\tilde{A}_{t_\infty}\|^2 < \infty.$$

(3) We consider again the process $A := h1_{[u, t_\infty]}$ as in (2) above, and the square integrable martingale $N := A - \tilde{A}$. Let M be an element of $\mathcal{M}^2_T(\mathbf{H})$. Using the property in Section 9.11, we may write for every stopping time v with values in $[v, t_\infty]$

$$\begin{aligned} E(M_v \otimes N_v) &= E\left(\int_{]0, v]} M_s \otimes dN_s\right) \\ &= E\left(\int_{]0, v]} M_{s^-} \otimes dN_s\right) + E\left(\int_{]0, v]} \Delta M_s \otimes dN_s\right) \end{aligned}$$

But since the Doléans measure of N is zero, the first integral is zero, and we obtain

$$E(M_v \otimes N_v) = E\left(\sum_{s \leqslant v} \Delta M_s \otimes \Delta N_s\right). \tag{10.1.4}$$

Since the only jumps of N are those of A, according to point (2), we see that N *is strongly orthogonal to every* M *in* $\mathcal{M}^2_T(\mathbf{H})$ *for which* $E\|\Delta M_u\|^2 = 0$.

We also have the following uniform approximation lemma.

Lemma *Let* $(M^n)_{n \in \mathbf{N}}$ *be a sequence of elements of* $\mathcal{M}^2_T(\mathbf{H})$, *which converges in* $\mathcal{M}^2_T(\mathbf{H})$ *toward* M. *Then there exists a subsequence* $(M^{n_k})_{k \in \mathbf{N}}$ *with the* *following property: for P-almost all* $\omega \in \Omega$, *the paths* $t \rightarrow M^{n_k}(t, \omega)$ *converge uniformly on* $[0, t_\infty]$ *to the path* $t \rightarrow M(t, \omega)$.

Proof As a consequence of the stopping theorem of Section 8.6 we have

$$E\left(\|M_u - M_u^n\|^2\right) \leqslant E\left(\|M_{t_\infty} - M_{t_\infty}^n\|^2\right)$$

for every n and every T-valued stopping time u. Therefore for the processes M^n we can use exactly the same argument as that for the processes Z^n in Section 2.5, so we omit the details here.

As a consequence of this lemma it is immediately seen, for example, that *the subspace of continuous square integrable martingales is a Hilbert subspace of* $\mathfrak{M}_T^2(\mathbf{H})$.

10.2 Meyer Processes of a Square Integrable Martingale

For a square integrable martingale M we write $\langle M, M\rangle$ (or for short $\langle M\rangle$) for the (increasing) Meyer process associated with the Doléans measure of the submartingale $\|M\|^2$. We may also define the $\mathbf{H}\hat{\otimes}_2\mathbf{H}$-valued process $\langle\!\langle M, M\rangle\!\rangle$ (or shortly $\langle\!\langle M\rangle\!\rangle$), which is the Meyer process associated with the Doléans msasure of $M \otimes M$.

In other words $\langle M, M\rangle$ is the unique predictable cadlag increasing process such that $\|M\|^2 - \langle M, M\rangle$ is a martingale.

In an analogous way $\langle\!\langle M, M\rangle\!\rangle$ is the unique predictable cadlag process with finite variation such that $M \otimes M - \langle\!\langle M, M\rangle\!\rangle$ is a martingale.

Proposition *If the square integrable martingales M and N are strongly orthogonal such that* $M_0 \otimes N_0 = 0$ a.s., *we have*

$$\langle M+N, M+N\rangle = \langle M, M\rangle + \langle N, N\rangle,$$
$$\langle\!\langle M+N, M+N\rangle\!\rangle = \langle\!\langle M, M\rangle\!\rangle + \langle\!\langle N, N\rangle\!\rangle.$$

Proof This follows immediately from the equality between Doléans measures:

$$d_{\|M+N\|^2} = d_{\|M\|^2} + d_{\|N\|^2},$$

which is a consequence of the strong orthogonality.

10.3 A Decomposition of Square Integrable Martingales

Lemma *For every cadlag square integrable martingale M and T-valued predictable stopping time u, the process* $\Delta M_u 1_{[u, t_\infty]}$ *is a cadlag square integrable martingale.*

Proof By using an announcing sequence (u_n) for u and the stopping theorem of Section 8.6, we obtain

$$E(M_u - M_{u_n} \mid \mathcal{F}_{u_n}) = 0 \qquad \text{for all} \quad n,$$

then

$$E(M_u - M_{u_n} \mid \mathcal{F}_{u^-}) = 0 \qquad \text{for all} \quad n,$$

and finally

$$E(M_u - M_{u^-} \mid \mathcal{F}_{u^-}) = 0.$$

We have already proved in Section 2.8 that $E\|\Delta M_u\|^2 < \infty$. The lemma is therefore an immediate consequence of Example (1) in Section 10.1 (see formula (10.1.1): in the present case $\tilde{A} = 0$).

Theorem *Every Hilbert-valued cadlag square integrable martingale M can be written under the following form*:

$$M = N + \sum_n \Delta M_{u(n)} 1_{[u(n), t_\infty]}, \tag{10.3.1}$$

where

(i) *the $u(n)$ are predictable stopping times with disjoint graphs*;

(ii) $M^n := \Delta M_{u(n)} 1_{[u(n), t_\infty]}$ *is a martingale on* $[0, t_\infty]$ *and the above convergence holds in* $\mathcal{M}_T^2(\mathbf{H})$;

(iii) $\langle N, N \rangle$ *is continuous.*

The processes N and $\sum_n \Delta M_{u(n)} 1_{[u(n), t_\infty]}$, entering the above decomposition, are determined uniquely up to P-equivalence by the conditions (i)–(iii). *Moreover, $\langle M, M \rangle = \langle N, N \rangle + \sum_n \langle M^n, M^n \rangle$, the convergence being uniform on $[0, t_m]$ for P-almost all ω.*

Proof Let us consider the Doléans measure α of the positive submartingale $\|M\|^2$. We decompose it as in the proof of Theorem 9.12:

$$\alpha = \mu + \sum_{n>0} \beta_n, \tag{10.3.2}$$

where μ is an admissible measure associated with a continuous Meyer process, and where β_n are carried by disjoint graphs $[u_n]$ of predictable stopping times. From the above lemma we see that the processes $\Delta M_{u(n)} 1_{[u(n), t_\infty]}$ are square integrable cadlag martingales. From Section 10.1, Example (3) we know that these martingales are strongly orthogonal. We have therefore

$$E(\Delta M_{u_n} \Delta M_{u_k}) = 0, \qquad \text{for} \quad k \neq n.$$

Moreover, since

$$\sum_n \|\Delta M_{u_n}\|^2 \leqslant [M]_{t_\infty},$$

the series $\sum_{n>0} \Delta M_{u_n} 1_{[u_n, t_\infty]}$ converges in $\mathfrak{M}_T^2(\mathbf{H})$ toward a martingale M^j such that $\sum_{n>0} \beta_n$ is the Doléans measure of $\|M^j\|^2$.

The existence of a jump of the martingale M on some predictable stopping time u with graph disjoint from the $[u_n]$s would contradict the definition of μ. Therefore $M - M^j$ has no predictable jump and is strongly orthogonal to M^j (see Section 10.1, Example (3)). As a consequence, the Doléans measure of $\|M\|^2$ is the sum of the Doléans measure of $\|M^j\|$ and $\|N\|^2 := \|M - M^j\|^2$ which means that μ is the Doléans measure of $\|N\|^2$. This establishes the decomposition (10.3.1).

The unicity of the decomposition $M = N + M^j$ with properties (i)–(iii) is trivial from the fact that the existence of another $M = N' + M^{j'}$ would imply that $N - N'$ has a continuous Meyer process and at the same time is the sum of jumps on predictable stopping times, which is contradictory (see Section 9).

To prove the last statement of the theorem we first notice that the strong orthogonality of the martingales $N, M^1, \ldots, M^n$ implies

$$\langle N + \sum_{k \leqslant n} M^k, N + \sum_{k \leqslant n} M^k \rangle = \langle N, N \rangle + \sum_{k \leqslant n} \langle M^k, M^k \rangle$$

as well as

$$E\left(\sum_{l=k}^{k+n} \langle M^l, M^l \rangle_{t_\infty} \right) = \sum_{l=k}^{k+n} E \|\Delta M_{u_l}\|^2.$$

Therefore

$$\lim_{k \to \infty} E \sum_{l \geqslant k} \langle M^l, M^l \rangle_{t_\infty} = 0 \text{ a.s.}$$

The uniform convergence on $[0, t_\infty]$ for P-almost all ω follows from the fact that the positive process $\sum_{l \geqslant k} \langle M^l, M^l \rangle$ is increasing.

Definition The process $\sum_n M^n$ in the decomposition of the theorem will be called the *pure jump part of* M, a martingale of the form $\sum_n h_n 1_{[u_n, t_\infty]}$, where the u_n are predictable stopping times and $E(h \mid \mathcal{F}_{u_n^-}) = 0$ will be called a *pure jump martingale*.

Remark There should be a distinction made between the pure jump part M^j of M and what is usually called (see [Mey-4]) the *purely discontinuous part* M. The purely discontinuous part M^d refers to another decomposition theorem that will be mentioned later (see Example 1 at the end of Section 10).

*-DOMINATING PROCESSES FOR SQUARE INTEGRABLE MARTINGALES

10.4 Statement of the Main Inequality

Theorem *Let M be an* **H***-valued square integrable martingale and M^j be its pure jump part as defined in Section* 10.3. *Then the increasing process $A := 1 + B$, where $B := 4\langle M\rangle + 4[M^j]$ is a *-dominating process for M. More precisely, for every $\mathcal{Q}$-simple process Y (with values in $\mathcal{L}(\mathbf{H};\mathbf{K})$,* **K***: Hilbert) and every stopping time u, we have*

$$E\left(\sup_{t<u}\left\|\int_{]0,t]} Y\,dM\right\|^2\right) \leqslant E\left(\int_{]0,u[}\|Y_s\|^2\,dB_s\right)$$

$$\leqslant E\left(A_{u^-}\int_{]0,u[}\|Y_s\|^2\,dA_s\right). \tag{10.4.1}$$

Proof The proof, which is elementary but far from being simple, will be given in several steps contained in Sections 10.5–10.7.

Examples 2 and 3 at the end of Section 10 show that the existence of a *-dominating process is not trivial.

Corollary (Stopped Doob Inequality) *Let M be a square integrable martingale and u be a stopping time. Then we have*

$$E\left(\sup_{t<u}\|M_t\|^2\right) \leqslant 4E(\langle M\rangle_{u^-} + [M^j]_{u^-}). \tag{10.4.2}$$

This statement is a trivial consequence of (10.4.1) by taking $Y = 1$.

Let us remark that the Doob inequality of Section 8.2(4) and the stopping theorem immediately give

$$E\left(\sup_{t\leqslant u}\|M_t\|^2\right) \leqslant 4E([M]_u) = 4E(\langle M\rangle_u).$$

If $\langle M\rangle$ has no jump on u, in particular if M is continuous or if u is totally inaccessible, we immediately derive in this particular case

$$E\left(\sup_{t<u}\|M_t\|^2\right) \leqslant 4E(\langle M\rangle_u) = 4E\langle M\rangle_{u^-}.$$

But Examples 2 and 3 at the end of Section 10 show that one cannot expect such an inequality in the general case.

We proceed now to the successive steps of the proof, which reproduces essentially our proof given in [MeP-6].

10.5 A Lemma on the Conditional Expectation

Lemma *We consider a probability space* $(\Omega, \mathcal{F}, P)$, *a sub-σ-algebra* $\mathcal{G}$ *of* $\mathcal{F}$, *an element* A *of* $\mathcal{F}$, *and the σ-algebra* $\mathcal{G}^*$ *generated by* $\mathcal{G}$ *and* A. *Let* Z *be an element of* $L^1_{\mathbf{H}}(\Omega, \mathcal{G}^*, P)$ *such that* $E(Z \mid \mathcal{G}) = 0$. *If we write* $B = \Omega \backslash A$, *we have*

(i) $E(1_A \mid \mathcal{G}) E(\|Z\|^2 1_A \mid \mathcal{G}) = E(1_B \mid \mathcal{G}) E(\|Z\|^2 1_B \mid \mathcal{G})$ a.e.

(ii) $E(1_B \|Z\|^2) = E(1_A E(\|Z\|^2 \mid \mathcal{G}))$.

Proof (i) We can write $Z = X 1_A + Y 1_B$, where X and Y belong to $L_1^{\mathbf{H}}(\Omega, \mathcal{G}, P)$. The property $E(Z \mid \mathcal{G}) = 0$ implies

$$XE(1_A \mid \mathcal{G}) = E(X 1_A \mid \mathcal{G}) = -E(Y 1_B \mid \mathcal{G}) = -YE(1_B \mid \mathcal{G}).$$

Then we also have

$$\begin{aligned} E(1_A \mid \mathcal{G}) E(\|Z\|^2 1_A \mid \mathcal{G}) &= E(1_A \mid \mathcal{G}) E(\|X\|^2 1_A \mid \mathcal{G}) \\ &= [E(1_A \mid \mathcal{G})]^2 \|X\|^2 = \|Y\|^2 [E(1_B \mid \mathcal{G})]^2 \\ &= E(1_B \mid \mathcal{G}) E(\|Z\|^2 1_B \mid \mathcal{G}). \end{aligned}$$

(ii) Elementary properties of conditional expectation give

$$\begin{aligned} E\{1_A E(\|Z\|^2 1_A \mid \mathcal{G})\} &= E\{E(1_A \mid \mathcal{G}) E(\|Z\|^2 1_A \mid \mathcal{G})\} \\ &= E\{E(1_B \mid \mathcal{G}) E(\|Z\|^2 1_B \mid \mathcal{G})\} \\ &= E\{1_B E(\|Z\|^2 1_B \mid \mathcal{G})\}. \end{aligned}$$

Then we also have

$$\begin{aligned} E\{1_A E(\|Z\|^2 \mid \mathcal{G})\} &= E\{1_A E(\|Z\|^2 1_A \mid \mathcal{G})\} + E\{1_A E(\|Z\|^2 1_B \mid \mathcal{G})\} \\ &= E\{1_B E(\|Z\|^2 1_B \mid \mathcal{G})\} + E\{1_A E(\|Z\|^2 1_B \mid \mathcal{G})\} \\ &= E(\|Z\|^2 1_B). \end{aligned}$$

10.6 Proposition

Proposition *Let* $\mathbf{H}$ *be a Hilbert space,* M *an* $\mathbf{H}$*-valued cadlag square integrable martingale,* u *a stopping time. Let* M^j *be the pure jump part of* M *and* $N := M - M^j$ *(see Section 10.3). Then there exists an* $\mathbf{H}$*-valued cadlag square integrable martingale* W *with the following properties*:

(i) $W 1_{[0, u[} = M 1_{[0, u[}$ *(this implies* $[W] 1_{[0, u[} = [M] 1_{[0, u[}$*)*;

(ii) *for every positive Borel function* ϕ *on* $[0, t_\infty]$ *we have*

$$\int \phi(s) d\langle W \rangle_s \leqslant \int \phi(s) d\langle M \rangle_s \text{ a.s.;}$$

(iii) *for every predictable real positive process* Y *we have*

$$E\left\{\int_{]0,u]} Y\,d[W]\right\} = E\left\{\int_{]0,u]} Y\,d\langle W\rangle\right\}$$
$$\leqslant E\left\{\int_{]0,u[} Y\,(d[W-N] + d\langle W\rangle)\right\}$$
$$\leqslant E\left\{\int_{]0,u[} Y\,(d[M^j] + d\langle M\rangle)\right\}.$$

Proof The proof of this proposition is given in two main steps. We prove it first for a pure jump martingale with only one jump (this is the core of the proof), and next we consider a general $M \in \mathfrak{M}_T^2(\mathbf{H})$.

(1) We assume $M = C1_{[v, t_\infty]}$, where v is a predictable stopping time and C is $\mathcal{F}_v$-measurable and $E(C \mid \mathcal{F}_{v^-}) = 0$ (see Example **3** at the end of Section 9). In this case, $N = 0$. We introduce the sets $A := \{\omega : v(\omega) < u(\omega)\}$ and $B := \Omega \backslash A$. For convenience we write $\mathcal{G}$ for $\mathcal{F}_{v^-}$; $\mathcal{G}^*$ is the σ-algebra generated by $\mathcal{G}$ and B. Writing

$$D := C1_B - E(C1_B \mid \mathcal{G}^*) = 1_B(C - E(C \mid \mathcal{G}^*)),$$

we obtain from the definition

$$E(D \mid \mathcal{G}^*) = 0 \qquad \text{and a fortiori} \qquad E(D \mid \mathcal{G}) = 0.$$

The process $D1_{[v, t_\infty]}$ is therefore a square integrable martingale. We set

$$W := M - D1_{[v, t_\infty]},$$

which can be still written

$$W = 1_A C 1_{[v, t_\infty]} + 1_B E(C \mid \mathcal{G}^*) 1_{[v, t_\infty]}.$$

The property (i) thus holds for W. We know (see (3) at the end of Section 9) that

$$\langle W\rangle = E(\|\Delta W_v\|^2 \mid \mathcal{F}_{v^-}) 1_{[v, t_\infty]}$$

and

$$\langle M\rangle = E(\|C\|^2 \mid \mathcal{F}_{v^-}) 1_{[v, t_\infty]}.$$

From the Jensen inequality, it follows that

$$E(1_A\|C\|^2 + 1_B\|E(C \mid \mathcal{G}^*)\|^2 \mid \mathcal{G}) \leqslant E(\|C\|^2 \mid \mathcal{G}),$$

and then,

$$E(\|\Delta W_v\|^2 \mid \mathcal{F}_{v^-}) \leqslant E(\|\Delta M_v\|^2 \mid \mathcal{F}_{v^-}),$$

which implies (ii) immediately. Let Y be a predictable real positive process.

Then the random variable Y_v is $\mathcal{F}_{v^-}$-measurable. We have

$$E\left(\int_{]0,\,u]} Y\,d[W]\right) = E\left(Y_v 1_A \|C\|^2\right) + E\left(Y_v 1_B \|E(C \mid \mathcal{G}^*)\|^2\right).$$

The first term is equal to

$$E\left(\int_{]0,\,u[} Y\,d[W]\right).$$

According to the lemma of Section 10.5, if we take $Z = (Y_v)^{1/2} E(C \mid \mathcal{G}^*)$, the second term is equal to

$$E\left(Y_v 1_A E\left\{\|E(C \mid \mathcal{G}^*)\|^2 \mid \mathcal{G}\right\}\right),$$

which is less than

$$E\left(\int_{]0,\,u[} Y\,d\langle W\rangle\right).$$

This proves property (iii).

(2) We consider now a general $M \in \mathfrak{M}_T^2(\mathbf{H})$ and write it $M = N + M^j$, according to the decomposition theorem of Section 10.3. More precisely, we have

$$M^j = \sum_n \Delta M_{v_n} 1_{[v_n,\, t_\infty]},$$

where (v_n) is a family of predictable stopping times with disjoint graphs. We write

$$M^n := \Delta M_{v_n} 1_{[v_n,\, t_\infty]}.$$

For $n \geqslant 1$ we set

$$A_n := \{v_n < u\}, \qquad B_n := \Omega - A_n,$$

and

$$W^n := 1_{A_n} \Delta M_{v_n} 1_{[v_n,\, t_\infty]} + 1_{B_n} E(\Delta M_{v_n} \mid \mathcal{G}_n^*) 1_{[v_n,\, t_\infty]}, \tag{10.6.1}$$

where $\mathcal{G}_n^*$ denotes the σ-algebra generated by $\mathcal{F}_{v_n^-}$ and A_n. The series $N + \sum_{n \geqslant 1} W^n$ is clearly convergent in $\mathfrak{M}_T^2(\mathbf{H})$, and we may set

$$W := N + \sum_{n \geqslant 1} W^n. \tag{10.6.2}$$

In view of the strong orthogonality of N and W^n, this implies (see the theorem in Section 10.3)

$$\langle W\rangle = \langle N\rangle + \sum_{n \geqslant 1} \langle W^n\rangle \qquad \text{(a.s. uniform convergence)},$$

and even more, from the one jump character of each $\langle W^n\rangle$

$$\int \phi\,d\langle W\rangle = \int \phi\,d\langle N\rangle + \sum_n \int \phi\,d\langle W^n\rangle \text{ a.s.} \tag{10.6.3}$$

for every Borel function ϕ on $[0, t_m]$. The analogous equality holds for M:

$$\int \phi \, d\langle M\rangle = \int \phi \, d\langle N\rangle + \sum_n \int \phi \, d\langle M^n\rangle. \tag{10.6.4}$$

Properties (i) and (ii) therefore follow immediately from formulas (10.6.1)–(10.6.3) and from the properties of W^n, established in the first part of the proof. From the definition we obtain immediately

$$\int \phi \, d[W - N] = \sum_n \int \phi \, d[W^n] \text{ a.s.} \tag{10.6.5}$$

for every Borel function ϕ on $[0, t_\infty]$. From the first part of the proof we may then derive, for every predictable positive process Y,

$$E\Big(\int_{]0,u]} Y \, d\langle W\rangle\Big) = E\Big(\int_{]0,u]} Y \, d\langle N\rangle\Big) + \sum_{n \geq 1} E\int_{]0,u]} Y \, d\langle W^n\rangle$$

$$\leq E\Big(\int_{]0,u]} Y \, d\langle N\rangle\Big) + \sum_{n \geq 1} \Big[E\Big(\int_{]0,u[} Y \, d\langle W^n\rangle\Big)$$

$$+ E\Big(\int_{]0,u[} Y \, d[W^n]\Big)\Big].$$

The continuity of $\langle N\rangle$ and relations (10.6.3) and (10.6.5) give

$$E\Big(\int_{]0,u]} Y \, d\langle W\rangle\Big) \leq E\Big(\int_{]0,u[} Y \, d\langle W\rangle\Big) + E\Big(\int_{]0,u[} Y \, d[W - N]\Big).$$

This is the first inequality in (ii). The second inequality is a straightforward consequence of the equality of $[M^j]$ and $[W - N]$ on $[0, u[$ and of (ii).

10.7 Proof of the Theorem in Section 10.4

Let W be associated with M as in the proposition in Section 10.6. From $W = M$ on $[0, u[$ we derive

$$\sup_{t<u} \Big\| \int_{]0,t]} Y \, dM \Big\|^2 \leq \sup_{t \leq u} \Big\| \int_{]0,t]} Y \, dW \Big\|^2$$

for every $\mathcal{C}$-simple process Y.

From the Doob inequality in Section 8.2, applied to the martingale $(\int Y \, dW)$ stopped at u, we obtain

$$E\Big(\sup_{t<u} \Big\| \int_{]0,t]} Y \, dM \Big\|^2\Big) \leq 4E\Big\| \int_{]0,u]} Y \, dW \Big\|^2.$$

Recalling once more that the Doléans measure of $\| W\|^2$ is defined by $[W]$ and $\langle W\rangle$ as well, we see that

$$E\Big(\sup_{t<u} \Big\| \int_{]0,t]} Y \, dM \Big\|^2\Big) \leq 4E\Big(\int_{]0,u]} \| Y\|^2 \, d[W]\Big) = 4E\Big(\int_{]0,u]} \| Y\|^2 \, d\langle W\rangle\Big).$$

Inequalities (iii) in the proposition in Section 10.6 then give the final result.

SEMIMARTINGALES AND SUMMABLE PROCESSES

The notion of semimartingales (at least for real processes) has been introduced and extensively studied by P. A. Meyer (see [Mey]).

The notion of p-summable processes was first introduced (under the name of "processus de répartition") in [Pel-3]. Prelocally p-summable processes are due to A. V. Kussmaul (see [Kus]).

10.8 Definitions

Definition 1 A Banach-valued process X on T is called a *semimartingale* if it can be written

$$X = M + V,$$

where M is a local martingale (see Section 9.16) and V is a process the paths of which have finite variation on every bounded interval $[0, t] \subset T$.

Definition 2 We say (cf. [Kus]) that an $\mathbf{H}$-valued process X is a *p-summable* process (with $p \geqslant 0$) if the mapping $A \rightarrow \int 1_A \, dX$ defined on the algebra $\mathcal{Q}$ can be extended in a measure with values in $L^p_{\mathbf{H}}(\Omega, \mathcal{F}, P)$, which is σ-additive for the topology of $L^p_{\mathbf{H}}(\Omega, \mathcal{F}, P)$. We say that X is a *prelocally p-summable process* (cf. Section 1.13) if there exists a sequence $(u(n))_{n>0}$ of stopping times such that

$$\lim_{n\to\infty} P([u(n) < \infty]) = 0$$

and such that for each integer n, $X^{u(n)}$ is a p-summable process (in [Kus] such a process is called locally p-summable).

10.9 Semimartingales as π^*-Processes

The following theorem extends in several ways a result of Kussmaul (see [Kus, Theorem 12.2]).

Theorem *Let* $\mathbf{H}$ *be a Hilbert space and* Z *an* $\mathbf{H}$*-valued cadlag adapted process. Let us consider the following statements*:

(i) Z *is a semimartingale*;
(ii) Z *is a* π^**-process*;
(iii) Z *is a* π*-process*;
(iv) Z *is prelocally* 2*-summable*;
(v) Z *is prelocally* 1*-summable*.

Then the following implications hold:

$$\text{(i)} \Rightarrow \text{(ii)} \Rightarrow \text{(iii)} \Rightarrow \text{(iv)} \Rightarrow \text{(v)}.$$

If $\mathbf{H}$ *is finite-dimensional, then all the statements are equivalent.*

Proof The implications (ii)$\Rightarrow$(iii) and (iv)$\Rightarrow$(v) are trivial from the definitions. We are left to prove (i)$\Rightarrow$(ii), (iii)$\Rightarrow$(iv), and in the finite-dimensional case, (v)$\Rightarrow$(i).

(1) (i)$\Rightarrow$(ii): From the theorem in Section 9.16 Z can be written $Z = M + V$, where M is a locally square integrable martingale and V a process with finite variation. Since V is a π^*-process (the proposition in Section 6.9), we have to prove that every locally square integrable martingale is a π^*-process. But we have proved in Section 10.4 that every square integrable martingale is a π^*-process. We may then apply the proposition in Section 6.9(4) which says that the π^*-property is "local."

(2) (iii)$\Rightarrow$(iv): Let $(u(n))$ be a localizing sequence of stopping times for the π-process X (see Section 2.9). If $\alpha^{u(n)}$ is a dominating measure associated with $u(n)$, the inequality

$$E\left\|\int 1_G\, dX^{u(n)}\right\|^2 \leqslant E\left(\int 1_G\, d\alpha^{u(n)}\right)$$

shows immediately the σ-additivity of $G \rightarrow \int 1_G\, dX^{u(n)}$.

(3) (v)$\Rightarrow$(i) when **H** is finite-dimensional: If Z is 1-summable, the **H**-valued mapping $G \rightarrow E(\int 1_G\, dZ)$ is a σ-additive measure on $\mathcal{P}$, which is nothing but the Doléans measure of Z. If **H** is finite-dimensional, this measure has bounded variation. One may therefore apply the theorem of Section 9.14, in which is shown the existence of a Meyer process V such that $Z - V$ is a martingale (having Doléans measure zero!). If Z is only prelocally bounded, we see from this argument that there exists an increasing sequence $(u(n))$ of stopping times converging to $+\infty$ such that for each n $Z^{u(n)}$ is a semimartingale. This completes the proof.

Remarks (1) The fact that there exist vector measures without finite variation shows immediately that the implications (v)$\Rightarrow$(i) and even (iv) $\Rightarrow$(i) are not true in general (see Example 4 below).

(2) The stochastic integral with respect to p-summable processes was built for the first time in [Pel-3], and later, for prelocally p-summable processes, in [Kus]. We have restricted ourselves until now to π-processes to avoid vector valued integration. The following chapter will be more specifically devoted to this point of view.

EXTENSIONS AND EXAMPLES

The details of the proofs are left as exercises.

1 A decomposition in $\mathfrak{M}_T^2(\mathbf{H})$ of a square integrable martingale M, different from the one in Section 10.3, can be given. One can indeed write

$$M = M^c + \sum_n (A_n - \tilde{A}_n) \qquad (\text{convergence in } \mathfrak{M}_T^2(\mathbf{H})),$$

where M^c is a continuous square integrable martingale, $A_n = \Delta M_{u_n} 1_{[u_n, t_\infty]}$, $\tilde{A}_n$ is the Meyer process of the Doléans measure of A_n, and (u_n) is a family of stopping times with disjoint graphs. (Remark that the martingales $A_n - \tilde{A}_n$ are strongly orthogonal. Show that the series $\sum_n (A_n - \tilde{A}_n)$ converges in $\mathfrak{M}_T^2(\mathbf{H})$ and, if M has no jump outside $\bigcup_n [u_n]$, that $M - \sum_n (A_n - \tilde{A}_n)$ is continuous.) For more details, see [Mey-4] or [Met-6].

2 We consider for Ω the two points set

$$\Omega := \{1, 2\},$$
$$\mathcal{F}_t := \{\emptyset, \Omega\} \quad \text{if} \quad t < 1,$$
$$\mathcal{F}_t := P(\Omega) \quad \text{if} \quad t \geqslant 1.$$

We define $u(\omega) = \omega$. Show that u is an accessible nonpredictable stopping time. If $P\{1\} = p_1$ and $P\{2\} = p_2 = 1 - p_1$, we define a martingale M by setting

$$M_t := \begin{cases} 0 & \text{if} \quad t < 1 \\ \dfrac{1}{p_1} 1_{\{1\}} - \dfrac{1}{p_2} 1_{\{2\}} & \text{if} \quad t \geqslant 1. \end{cases}$$

One can easily evaluate

$$E\Big(\sup_{s<u} |M_s|^2\Big) = \frac{1}{p_2} \qquad \text{and} \qquad E(\langle M \rangle_{u^-}) = \frac{1}{p_1}.$$

This provides us with a counterexample to an inequality of the type

$$E\Big(\sup_{s<u} |M_s|^2\Big) \leqslant CE(\langle M \rangle_{u^-})$$

for some constant C.

3 The following example shows the impossibility of a formula of the type

$$E\Big(\sup_{s<u} |M_s|^2\Big) < CE([M]_{u^-}).$$

Take

$$\Omega := \{1, 2, \ldots, n\}.$$

For $t \in [k, k+1[$, the σ-algebra $\mathcal{F}_t$ is generated by the atoms $\{j\}_{j \leqslant \inf(k, n)}$. Let q be a positive number with $0 < q < 1$:

$$P\{1\} = q, \quad P\{2\} = (1-q)q, \ldots, P(n-1) = (1-q)^{n-2} q,$$
$$P(n) = (1-q)^{n-1}.$$

The martingale M is defined in the following way: $M_t(\omega)$ is constant for

$t \in [k, k+1[$ and

$$M_0 = 0, \qquad M_{k+1} = M_k - \frac{1}{q} 1_{\{k+1\}} + \frac{1}{(1-q)} 1_{\{k+2,\ldots,n\}}.$$

Show that the stopping time u, defined by $u(\omega) = \omega$, is accessible but not predictable. Compute $E[M]_{u^-}$ and $E(\sup_{0 \leqslant t < u} |M_t|^2)$ and show that

$$\lim_{u \to 0} \frac{E[M]_{u^-}}{E\left(\sup_{0 \leqslant t < u} |M_t|^2\right)} = \frac{1}{n-1}.$$

4 Take $\mathbf{H} := L^2[0,1]$, $T = [0,1]$, and consider the $\mathbf{H}$-valued deterministic process ("spectral process")

$$X_t(\omega) = 1_{]0,t]} \in L^2[0,1] \qquad \text{for all} \quad \omega.$$

Show that X verifies the inequality of π^*-processes for real processes Y, with dominating process $Q_t = t$, without being a semimartingale.

5 *Emery Topology on Semimartingales* For the case in which $\mathbf{K}$ is a finite-dimensional space, Emery (cf. [Eme-2, 3]) defined a topology on the space of $\mathbf{K}$-valued semimartingales (cf. the theorem of Section 10.9), which we compare with the *theorem introduced in Section 7.3*. For every $\mathbf{K}$-valued semimartingale Z (or equivalently, π-process) set

$$G(Z) := \inf\{P(u < t_m) + E(A_{t_m} + [M]_{t_m})\},$$

this infimum being taken for all stopping times u and all decompositions $Z^u = V + M$ of Z^u (see Section 1.13), where M is a square integrable martingale and V is a process with finite variation A.

The Emery topology can be defined with G exactly as the topology was defined with F in Section 7.3. To show that this topology is stronger than the one defined in Section 7.3 when we consider $\mathbf{K}$-valued π^*-processes, first set $a := [G(Z)]^{1/2} < 1$; let u be a stopping time and $Z^u := V + M$ with $P(u < t_m) \leqslant a$, $E([M]_{t_m}) \leqslant a^2$, and $E(A_{t_m}) \leqslant a^2$, where A is the total variation of V and M is a square integrable martingale. Set $B := 2A + 8[M] + 8\langle M \rangle$. Remember (see Section 10.4) that the process $(1 + (1/a^2) B)$ is $*$-dominating for $(1/a)Z^u$. Then $(a + (1/a)B)$ is $*$-dominating for Z^u and

$$F(Z) \leqslant 2E(1 \wedge (a + (1/a)B)) \leqslant 2[a + 18a] \leqslant 38[G(Z)]^{1/2}.$$

Both topologies are actually easily seen to be equivalent. But note that G would not be defined for the process X in Exercise 4 above in which $\mathbf{K}$ is infinite-dimensional.

Let us mention that Emery has given in [Eme-3] other equivalent, interesting definitions of his topology on the space of real semimartingales.

11 INEQUALITIES

INEQUALITIES FOR MARTINGALES

11.1 Theorem

Theorem *Let $(\Omega, \mathfrak{F}, P, (\mathfrak{F}_t)_{t \in T})$ be a stochastic basis and X be a martingale with respect to this basis such that $X_0 = 0$. For convenience of notation we suppose that $T = [0, 1]$.*

We denote by $\mathfrak{U}$ the space of all processes Y such that

$$Y = \sum_{k=1}^{n-1} 1_{]u(2k),\, u(2k+1)]}$$

where $(u(j))_{1 \leqslant j \leqslant 2n}$ is an increasing family of stopping times. We define

$$\mathbf{N}_1(X) := E\Big(\sup_{t \in T} |X_t|\Big),$$

$$\mathbf{N}_1^-(X) := \liminf_{n \uparrow \infty} E\left[\left\{\sum_{k=1}^{2^n} (X_{k2^{-n}} - X_{(k-1)2^{-n}})^2\right\}^{1/2}\right],$$

$$\mathbf{N}_1^+(X) := \limsup_{n \uparrow \infty} E\left[\left\{\sum_{k=1}^{2^n} (X_{k2^{-n}} - X_{(k-1)2^{-n}})^2\right\}^{1/2}\right],$$

$$\mathbf{N}_1'(X) := \sup_{Y \in \mathfrak{U}} E\left|\int_{]0,1]} Y\,dX\right|.$$

Then these four seminorms are uniformly equivalent on the space of all real cadlag martingales with $X_0 = 0$. More precisely, we have

$$\mathbf{N}_1^+(X) \leqslant 2\sqrt{2}\,\mathbf{N}_1(X),$$
$$\mathbf{N}_1'(X) \leqslant 4\mathbf{N}_1^-(X),$$
$$\mathbf{N}_1(X) \leqslant 4\mathbf{N}_1'(X),$$
$$\mathbf{N}_1^-(X) \leqslant \mathbf{N}_1^+(X).$$

Proof The last inequality is obvious; the three others are proved, respectively, in Sections 11.2, 11.4, and 11.5. The reader will notice that those proofs do not use the materials of Section 9.

11.2 Proof of $\mathbf{N}_1^+(X) \leqslant 2\sqrt{2}\,\mathbf{N}_1(X)$

It is sufficient to prove this inequality for a martingale $(\Omega, \mathfrak{F}, P, (\mathfrak{F}_k, X_k)_{0 \leqslant k \leqslant n})$ (i.e., card$(T) < +\infty$). Moreover, for every positive number a we can put $X_n^a = X_n 1_{\{|X_n| \leqslant a\}}$ and $X_k^a = E(X_n^a \mid \mathfrak{F}_k)$; according to the Lebesgue theorem, we have $\mathbf{N}_1^+(X) = \lim_{a\to\infty} N_1^+(X^a)$ and $\mathbf{N}_1(X) = \lim_{a\to\infty} \mathbf{N}_1(X^a)$; thus it is sufficient to consider the case in which X is a square integrable martingale.

For such a martingale $(X_k)_{0\leqslant k\leqslant n}$ we put $a := \mathbf{N}_1(X)$ and $B_k := \sup\{a, \sup_{j\leqslant k}|X_j|\}$, and we have $a \leqslant E(B_n) \leqslant 2a$ and $E(B_n - B_0) \leqslant a$.

Let V be the quadratic variation of X, i.e.,

$$V_k := \left[\sum_{j=1}^{k-1} (X_{j+1} - X_j)^2\right].$$

For every process U (with $U = A$, Z, etc.) we put $\Delta_k U := U_{k+1} - U_k$. Moreover, we define the processes A and Z by $A_k := 1/B_k$ and $Z_k := \sum_{j=0}^{k-1} X_j \Delta_j X = \frac{1}{2}(X_k^2 - V_k)$.

In the sequel we write $\sum$ instead of $\sum_{k=0}^{n-1}$. According to the Schwarz inequality, we have

$$\begin{aligned} E\left[(V_n)^{1/2}\right] &\leqslant \left[E(B_n)\right]^{1/2}\left[E(A_n V_n)\right]^{1/2} \\ &\leqslant (2a)^{1/2}\left[E(A_n X_n^2) - 2E(A_n Z_n)\right]^{1/2} \\ &\leqslant (2a)^{1/2}\left[a - 2E(A_n Z_n)\right]^{1/2}. \end{aligned}$$

We have

$$\begin{aligned} A_n Z_n &= \sum \Delta_k(AZ) \\ &= \sum (\Delta_k A)(\Delta_k Z) + \sum Z_k(\Delta_k A) + \sum A_k(\Delta_k Z). \end{aligned}$$

But Z is a martingale and the expectation of the third sum above is equal to zero; moreover, since A is decreasing, we have

$$-Z_k(\Delta_k A) = \tfrac{1}{2}\left(V_k(\Delta_k A) - X_k^2(\Delta_k A)\right) \leqslant -\tfrac{1}{2}X_k^2(\Delta_k A),$$

which implies

$$\begin{aligned} -E(A_n Z_n) &\leqslant -E\left[\sum (\Delta_k A)(\Delta_k Z)\right] - \frac{1}{2}E\left(\sum X_k^2(\Delta_k A)\right) \\ &\leqslant E\sum\left|\left(X_k(\Delta_k X) + \frac{1}{2}X_k^2\right)(\Delta_k A)\right| \\ &\leqslant E\sum\left(\frac{1}{2}|X_k|^2 + |X_k X_{k+1}|\right)\frac{\Delta_k B}{B_k B_{k+1}} \\ &\leqslant \frac{3}{2}E\sum \Delta_k B \leqslant \frac{3}{2}E(B_n - B_0) \leqslant \frac{3a}{2}. \end{aligned}$$

At last we obtain

$$\begin{aligned} E\left[(V_n)^{1/2}\right] &\leqslant (2a)^{1/2}\left[a + 3a\right]^{1/2} \\ &\leqslant a2\sqrt{2}. \end{aligned}$$

11.3 Fefferman Inequality

Proposition *Let* $(\Omega, \mathcal{F}, P, (\mathcal{F}_k)_{0 \leqslant k \leqslant n})$ *be a stochastic basis and* $(A_k, B_k)_{0 \leqslant k \leqslant n}$ *an associated family of pairs of random variables; we assume that for every integer* k, A_k *is* $\mathcal{F}_k$*-measurable. We define*

$$D := \operatorname*{ess\,sup}_{k,\,\omega} E\left(\sum_{j=k}^{n} |B_j|^2 \mid \mathcal{F}_k \right) = \sup_k \left\| E\left(\sum_{j=k}^{n} |B_j|^2 \mid \mathcal{F}_k \right) \right\|_\infty .$$

Then we have

$$E\left(\sum_{k=0}^{n} |A_k||B_k| \right) \leqslant (2D)^{1/2} E\left[\left(\sum_{k=0}^{n} |A_k|^2 \right)^{1/2} \right].$$

Proof It is sufficient to consider the case in which the random variables A_k and B_k are positive; but of course the inequality holds for Banach space valued random variables.

Let X and Y be the increasing processes defined respectively by $X_{-1} := Y_{-1} := 0$ and for $k \geqslant 0$ by

$$X_k := \sum_{j=0}^{k} A_j^2 \qquad \text{and} \qquad Y_k := (X_k)^{1/2} + (X_{k-1})^{1/2},$$

which implies $(Y_k)^{-1} A_k^2 = (X_k)^{1/2} - (X_{k-1})^{1/2}$.

According to the Schwarz inequality, we have

$$\begin{aligned} \sum_{k=0}^{n} A_k B_k &\leqslant \left(\sum_{k=0}^{n} (Y_k)^{-1} A_k^2 \right)^{1/2} \left(\sum_{k=0}^{n} B_k^2 Y_k \right)^{1/2} \\ &\leqslant (X_n)^{1/4} \left\{ \sum_{k=0}^{n} \left[(Y_k - Y_{k-1}) \sum_{j=k}^{n} B_j^2 \right] \right\}^{1/2} . \end{aligned}$$

Then the Schwarz inequality implies that

$$\left[E\left(\sum_{k=0}^{n} A_k B_k \right) \right]^2 \leqslant E\left[(X_n)^{1/2} \right] E\left\{ \sum_{k=0}^{n} \left[(Y_k - Y_{k-1}) \sum_{j=k}^{n} B_j^2 \right] \right\}.$$

The last expectation above is also equal to

$$\begin{aligned} E\left\{ \sum_{k=0}^{n} \left[(Y_k - Y_{k-1}) E\left(\sum_{j=k}^{n} B_j^2 \mid \mathcal{F}_k \right) \right] \right\} &\leqslant DE\left\{ \sum_{k=0}^{n} (Y_k - Y_{k-1}) \right\} \\ &\leqslant DE(Y_n) \leqslant 2DE\left[(X_n)^{1/2} \right], \end{aligned}$$

which implies

$$\left[E\left(\sum_{k=0}^{n} A_k B_k \right) \right]^2 \leqslant 2D\left(E\left[(X_n)^{1/2} \right] \right)^2 .$$

11.4 Proof of $\mathbf{N}_1'(X) \leqslant 4\mathbf{N}_1^-(X)$

We define

$$\mathfrak{L} := \{Z : Z \in L_\infty(\Omega, \mathfrak{F}, P), \|Z\|_\infty \leqslant 1\}.$$

Let X be a martingale, Z_1 an element of $\mathfrak{L}$, and Y an element of $\mathfrak{U}$ with $Y = \sum_{k=1}^n Y_k 1_{]u(k-1), u(k)]}$, where $(u(k))_{0 \leqslant k \leqslant n}$ is an increasing family of simple stopping times, Y_k is $\mathfrak{F}_{u(k-1)}$-measurable, and $Y_k(\omega) = 0$ or 1 for all $\omega \in \Omega$. Let Z and W be the martingales defined respectively by

$$Z_t := E(Z_1 \mid \mathfrak{F}_t) \qquad \text{and} \qquad W_t := \int_{]0,t]} Y_s \, dX_s.$$

For every process V we write as usual

$$\Delta_k V := V_{u(k)} - V_{u(k-1)},$$

and we write $\sum$ instead of $\sum_{k=1}^n$ and $\bar{Z}_k$ instead of $Z_{u(k)}$. We have

$$\begin{aligned} E(Z_1 W_1) &= E\left\{\sum \Delta_k (ZW)\right\} \\ &= E\left\{\sum (\Delta_k Z)(\Delta_k W)\right\} \\ &= E\left\{\sum Y_{u(k)} (\Delta_k Z)(\Delta_k X)\right\}; \end{aligned}$$

according to the Fefferman inequality, we obtain

$$E(Z_1 W_1) \leqslant \sqrt{2}\, E\left\{\left[\sum Y_k^2 (\Delta_k X)^2\right]^{1/2}\right\}\left\{\sup_k \left\| E\left(\sum_{j=k}^n (\Delta_j Z)^2 \mid \mathfrak{F}_{u(k)}\right)\right\|_\infty\right\}^{1/2}$$

But

$$\sum_{j=k}^n (\Delta_j Z)^2 = Z_1^2 - \bar{Z}_k^2 - 2 \sum_{j=k+1}^{2n} \bar{Z}_{j-1} \Delta_j Z + \left(\bar{Z}_k - \bar{Z}_{k-1}\right)^2$$

and

$$E\left\{\sum_{j=k}^n (\Delta_j Z)^2 \mid \mathfrak{F}_{u(k)}\right\} = E\left\{\left(Z_1 - \bar{Z}_k\right)^2 + \left(\bar{Z}_k - \bar{Z}_{k-1}\right)^2 \mid \mathfrak{F}_{u(k)}\right\} \leqslant 8,$$

which implies that $E(Z_1 W_1) \leqslant 4\mathbf{N}_1^-(X)$.

Now

$$\begin{aligned} \mathbf{N}_1'(X) &= \sup_{y \in \mathfrak{U}} E\left(\left|\int Y \, dX\right|\right) \\ &= \sup_{y \in \mathfrak{U},\, Z_1 \in \mathfrak{L}} E(Z_1 W_1) \leqslant 4\mathbf{N}_1^-(X). \end{aligned}$$

11.5 Proof of $\mathbf{N}_1(X) \leqslant 4\mathbf{N}_1'(X)$

If Z is a random variable, we put $Z^+ = \sup(Z, 0)$ and $Z^- = \sup(-Z, 0)$.

(1) Let $(\Omega, \mathcal{F}, P, (\mathcal{F}_k, X_k)_{0 \leqslant k \leqslant n})$ be a martingale such that $X_0 = 0$ and $X_{k+1}(\omega) < X_k(\omega)$ implies $X_j(\omega) = X_{k+1}(\omega)$ for $k+1 \leqslant j \leqslant n$: in other words, the sample function $k \curvearrowright X_k(\omega)$ is "fixed" after its first "going down."

We put $A(n) := \emptyset$, and if $0 \leqslant k < n$, $A(k) := \{\omega : X_{k+1}(\omega) < X_k(\omega)\}$; for every k, we define $B(k) := \bigcup_{j<k} A(j)$ and $C(k) := \Omega \backslash B(k)$ and $\Delta_k X := X_{k+1} - X_k$.

For every process $Y = \sum_k U_k 1_{]k, k+1]}$ we put $\int_0^j Y\,dX = \sum_{k=0}^{j-1} U_k \Delta_k X$.

Now we define recursively the sequence $(Y_k)_{0 \leqslant k \leqslant n}$ of (predictable!) processes and the associated sequence $(Z_k)_{0 \leqslant k \leqslant n}$ of random variables as follows:

$$Y_0 := 1, \qquad Z_k := \int_0^k Y_k\,dX;$$

$$Y_{k+1} := Y_k$$

if

$$E\left(1_{A(k)}\{3(Z_k + \Delta_k X)^+ + (Z_k + \Delta_k X)^- - Z_k\}\right) \geqslant 0$$

and $Y_{k+1} := Y_k - 1_{]k, k+1]}$ in the other case.

We put

$$S_k = 1_{C(k)} \sup_j \int_0^j Y_k\,dX + 1_{B(k)}(3Z_k^+ + Z_k^-)$$

and

$$s_k = E(S_k).$$

In (2) below we prove that $s_{k+1} \geqslant s_k$ for every integer k; this implies $s_n \geqslant s_0$, i.e.,

$$E(3Z_n^+ + Z_n^-) \geqslant E\left(\sup_k X_k\right).$$

($(Z_k)_{1 \leqslant k \leqslant n}$ is a Burkholder transform of $(X_k)_{1 \leqslant k \leqslant n}$ as defined in [Bur]). But $E(Z_n^+) = E(Z_n^-) = \frac{1}{2} E(|Z_n|)$ and $2E(|Z_n|) \geqslant E(\sup_k X_k)$.

(2) Now we fix k and prove $s_{k+1} \geqslant s_k$. For convenience of notation, we put $W := \int_0^{k+1} Y_k\,dX = Z_k + \Delta_k X$ and $D := C(k+1)$; moreover we write Z, ΔX, A, B, C instead of Z_k, $\Delta_k X$, $A(k)$, $B(k)$, $C(k)$, respectively. On B $X_k = X_n$ and on A we have $S_k = \sup_j \int_0^j Y_k\,dX = Z$ and $S_{k+1} = 3Z_{k+1}^+ + Z_{k+1}^-$. First we consider the case in which $Y_{k+1} := Y_k$, i.e., $Z_{k+1} = Z_k + \Delta X = W$ and

$$E\left[1_A(3W^+ + W^- - Z)\right] \geqslant 0.$$

In this case $S_{k+1} - S_k = 1_A(3W^+ + W^- - Z)$ and $s_{k+1} \geqslant s_k$. Now we consider the case in which

$$Y_{k+1} := Y_k - 1_{]k,k+1]}$$

and

$$E[1_A(3W^+ + W^- - Z)] \leqslant 0.$$

In this case $1_B S_k = 1_B S_{k+1}$, $Z_{k+1} = Z_k = Z$,

$$1_A(S_{k+1} - S_k) = 1_A(3Z^+ - Z) = 1_A 2Z$$
$$1_D(S_{k+1} - S_k) = 1_D(-\Delta X) \leqslant 0.$$

Thus we have

$$\Delta s := s_{k+1} - s_k = E(1_A 2Z) - E(1_D \Delta X).$$

Because of the martingality of X, this implies

$$\Delta s = E(1_A 2Z) + E(1_A \Delta X) = E(1_A W) + E(1_A Z)$$
$$= E(1_A W^+) - E(1_A W^-) + E(1_A Z).$$

But

$$-E(1_A W^-) \geqslant E(1_A 3W^+) - E(1_A Z),$$

which implies

$$\Delta s \geqslant E(1_A 4W^+) \geqslant 0.$$

(3) Now we consider a real cadlag martingale $(\Omega, \mathcal{F}, P, (\mathcal{F}_t, X_t)_{t\in T})$ with $X_0 = 0$. Let (a, ϵ) be two real numbers with $a > 1$ and $\epsilon > 0$; let $(u(n))_{n>0}$ be the sequence of stopping times defined by $u(n) = \inf\{t : X_t > a^n\epsilon\}$. We have

$$E\Big(\sup_{t\in T} X_t\Big) \leqslant a\epsilon + a\Big[\sup_n E\Big\{\sup_{0\leqslant k\leqslant n} X_{u(k)}\Big\}\Big].$$

But the martingale $(X_{u(k)})_{1\leqslant k\leqslant n}$ fulfills the conditions given in (1) above, and we have

$$E\Big(\sup_{0\leqslant k\leqslant n} X_{u(k)}\Big) \leqslant 2\mathbf{N}'_1(X).$$

This implies

$$E\Big(\sup_{t\in T} X_t\Big) \leqslant a\epsilon + 2a\mathbf{N}'_1(X)$$
$$\leqslant 2\mathbf{N}'_1(X) \qquad \text{if} \quad \epsilon\downarrow 0 \quad \text{and} \quad a\downarrow 1.$$

We can prove in the same way that

$$E\Big(\sup_{t\in T}(-X_t)\Big) \leqslant 2\mathbf{N}'_1(X)$$

and

$$\mathbf{N}_1(X) \leqslant 4\mathbf{N}'_1(X).$$

THE SPACES H_1 AND BMO

In the sequel we consider a stochastic basis $(\Omega, \mathcal{F}, P, (\mathcal{F}_t)_{t\in T})$ with $T = [0, 1]$.

11.6 The Space H_1

One calls $\mathbf{H}_1$ (with respect to a stochastic basis) the space of all the real martingales X such that $X_0 = 0$ and such that $\mathbf{N}_1^+(X) < +\infty$.

This space $\mathbf{H}_1$ is a Banach space for each one of the four equivalent norms $\mathbf{N}_1$, $\mathbf{N}_1^+$, $\mathbf{N}_1^-$, and $\mathbf{N}_1'$: the completeness is obvious for the norm $\mathbf{N}_1$. We saw in Section 10.9 that a martingale is a π-process: thus if X belongs to $\mathbf{H}_1$, the stochastic integral $\int Y\,dX$ is defined for any locally bounded (see Section 1.12) predictable process Y. Moreover $\mathbf{N}_1^+(X) = \mathbf{N}_1^-(X) = E([X]^{1/2})$.

Actually if X belongs to $\mathbf{H}_1$, the mapping $Y \rightsquigarrow \int Y\,dX$ is a linear mapping from $\mathcal{E}$ (see Section 2.2) into $\mathbf{L}_1(\Omega, \mathcal{F}, P)$: thus the stochastic integral $\int Y\,dX$ can be considered as a classical vector integral of the real process Y, considered as a real function on $(\Omega \times T)$, with values in the Banach space $\mathbf{L}_1$. In this setting, $\mathbf{N}_1'(\int Y\,dX)$ is exactly the classical "seminorm of the semivariation" of Y as considered by Bartle in [Bar] and others. Moreover (see Chapter 5), if X belongs to $\mathbf{H}_1$, the mapping $\mathcal{A} \rightsquigarrow \int 1_A\,dX$ defined on the algebra $\mathcal{A}$ (see Section 1.7) can be extended into a vector measure defined on the σ-algebra of predictable sets, with values in the Banach space $\mathbf{L}_1$ and σ-additive for the usual topology of $\mathbf{L}_1$.

Thus we can use all the classical results in vector integration. For example, the family of random variables $(\int Y\,dX)$ for all the processes Y such that $\mathbf{N}_1'(\int Y\,dX) \leqslant 1$ is uniformly equi-integrable (see [Mey-4]): this is a classical property of the vector integral (see [BDS]).

This construction of the stochastic integral as a classical vector integral was introduced in [Pel-3] and will be generalized in Chapter 5.

11.7 Seminorm N_∞ and Space BMO

Notation and Definition Let X be a real martingale with $X_0 = 0$. We denote by $\mathcal{T}$ the set of all the stopping times and we put

$$\mathbf{N}_\infty(X) := \sup_{u\in\mathcal{T}} \left\{ \| E\left[(X_1 - X_u)^2 | \mathcal{F}_u\right] + (X_u - X_{u^-})^2 \|_\infty \right\}^{1/2}.$$

One calls BMO the space of all the real martingales X such that $X_0 = 0$ and $\mathbf{N}_\infty(X) < +\infty$. It is easily seen that this space BMO is a Banach space for the norm $\mathbf{N}_\infty$ (consider a sequence $(X_n)_{n>0}$ such that $\mathbf{N}_\infty(X_n - X_{n+1})$

$\leqslant 2^{-n}$ and use the Borel–Cantelli lemma). Moreover, if Y is a real predictable process such that $\sup_{t,\omega}|Y_t(\omega)| \leqslant 1$, we have $\mathbf{N}_\infty(\int Y\,dX) \leqslant \mathbf{N}_\infty(X)$.

11.8 BMO is the dual of $\mathbf{H}_1$

Theorem *For every pair (X, Z) of martingales with $X \in \mathbf{H}_1$ and $Z \in$ BMO we have*

$$E\left(\int_{]0,1]} |d[X,Z]|\right) \leqslant 2\mathbf{N}_1^+(X)\mathbf{N}_\infty(Z) < +\infty,$$

and we can define the bilinear duality form

$$\langle X, Z\rangle := E\left(\int_{]0,1]} d[X,Z]\right).$$

Then the mapping $Z \mapsto \langle \cdot, Z\rangle$ defines an isomorphism from BMO *onto* $(\mathbf{H}_1)'$ *the topological dual of* $\mathbf{H}_1$.

Proof (1) Let $\epsilon > 0$. We define recursively the sequence of stopping times

$$u(0) := 0,$$
$$u(k+1) := \inf\{t : t \geqslant u(k), |X_t - X_{u(k)}| + |Z_t - Z_{u(k)}| > \epsilon\}.$$

We put $\Delta_k X = X_{u(k+1)} - X_{u(k)}$ and do the same for $\Delta_k Z$, etc. We have

$$(\Delta_{k-1} Z)^2 \leqslant 2(Z_{u(k)} - Z_{u(k)-})^2 + 2\epsilon^2.$$

This inequality and the Fefferman inequality imply (see the proof of Section 11.4)

$$E\left\{\sum_{k\geqslant 0} (|\Delta_k X|\,|\Delta_k Z|)\right\} \leqslant 2E\left\{\left[\sum_{k\geqslant 0} (\Delta_k X)^2\right]^{1/2}\right\}(\mathbf{N}_\infty(Z) + \epsilon).$$

According to Section 4.1, this implies

$$E\left(\int_{]0,1]} |d[X,Z]|\right) \leqslant 2\mathbf{N}_1^+(X)\mathbf{N}_\infty(Z)$$

(actually, it can be proved that

$$E\left(\int_{]0,1]} |d[X,Z]|\right) \leqslant \sqrt{2}\,\mathbf{N}_1^+(X)\mathbf{N}_\infty(Z),$$

which is the classical Fefferman inequality).

This shows that $\langle X, Z\rangle := E(\int_{]0,1]} d[X,Z])$ defines a continuous bilinear mapping on $(\mathbf{H}_1 \times \text{BM0})$.

(2) $\mathfrak{M}^2$ is dense in $\mathbf{H}_1$. We consider the space $\mathfrak{M}^2$ of square integrable martingales (see Section 10.1): $\mathfrak{M}^2$ is a Hilbert space for the scalar product

$\langle M, N\rangle = E(M_1 N_1)$ which coincides on $(\mathfrak{M}^2 \times \mathrm{BMO})$ with the bilinear mapping $\langle\cdot,\cdot\rangle$ defined above (of course $\mathfrak{M}^2$ is included in $\mathbf{H}_1$).

Moreover, in the sequel we denote by $\overline{\mathfrak{M}}^2$ the adherence of $\mathfrak{M}^2$ in $\mathbf{H}_1$. Let X be an element of $\mathbf{H}_1$ and $(u(n))_{n>0}$ be the sequence of stopping times defined by

$$u(n) := \inf\{t : |X_t| > n\}.$$

If we denote by X^n the martingale X stopped at the stopping time $u(n)$, we have clearly $\lim_{n\to\infty} \mathbf{N}_1(X - X^n) = 0$. Then it is sufficient to prove that X^n belongs to $\overline{\mathfrak{M}}^2$. We put $M := X^n$, and we look at the proof of Section 9.16; in this proof M is shown to be equal to $(A - Z) + (B - A)$, where $(A - Z)$ belongs to $\mathfrak{M}^2$ and $(B - A) = \lim_{\mathbf{H}_1}(B^n - A^n)$ belongs to $\overline{\mathfrak{M}}^2$ with

$$B_1^n := D 1_{|D| \leqslant n} \qquad \text{and} \qquad D := M_{u(n)} - M_{u(n)^-} = X_{u(n)} - X_{u(n)^-}.$$

(3) In the sequel we denote by K the subset of all the elements X of $\mathfrak{M}^2$ such that $\mathbf{N}_1^-(X) \leqslant 1$. Now we prove that for every element X of $\mathfrak{M}^2$ and for every stopping time u we have

$$b \leqslant 2 \sup_{Z \in K} \langle X, Z\rangle \qquad \text{where} \quad b := \|X_u - X_{u^-}\|_\infty.$$

Let $\epsilon > 0$. Let t be an element of T such that

$$b - \epsilon \leqslant \lim_{s \uparrow t} \|(X_u - X_{u^-}) 1_{]s,t]}\|_\infty$$

(if such an element t of T did not exist, $\|X_u - X_{u^-}\|_\infty \leqslant b - \epsilon$). We define

$$A(s) := \{\omega : b - \epsilon \leqslant 1_{]s,t]}(\omega) \|X_u - X_{u^-}\|(\omega)\},$$

$$a(s) := P[A(s)],$$

$$Z_1^s := \frac{1}{2a(s)b} 1_{A(s)} (X_u - X_{u^-}) 1_{]u,1]}$$

$Z^s := Z_1^s - Z_2^s$, where Z_2^s is the Meyer process associated with Z_1^s.

For every element s of T with $s < t$, Z^s is a (right continuous) martingale. By left continuity, there exists $s < t$ such that

$$\mathbf{N}_1^-(Z^s) \leqslant \epsilon + \frac{1}{2a(s)b} 2E\{1_{A(s)}|X_u - X_{u^-}|\} \leqslant \epsilon + 1$$

and

$$\langle X, Z^s\rangle \geqslant -\epsilon + \frac{1}{2a(s)b} E\{1_{A(s)}(X_u - X_{u^-})^2\} \geqslant -\epsilon + \frac{(b-\epsilon)^2}{2b},$$

which implies

$$b \leqslant 4\epsilon + 2 \sup_{\mathbf{N}_1^-(Z) \leqslant 1 + \epsilon} \langle X, Z\rangle.$$

(4) Let X be an element of $\mathfrak{M}^2$ and u be a stopping time. We define

$$\bar{b} := \left\{\text{ess sup } E\left[(X_1 - X_u)^2 | \mathcal{F}_u\right]\right\}^{1/2} \quad \text{and} \quad b := \bar{b}(1-\epsilon),$$

$$A := \left\{\omega : b^2 \leqslant E\left[(X_1 - X_u)^2 | \mathcal{F}_u\right](\omega)\right\},$$

$$a := P(A),$$

$$Z_t := (1/ab)(X_t - X_u)1_A 1_{]u,1]},$$

$$f := [Z, Z]_1.$$

According to the Schwarz inequality, we have

$$\mathbf{N}_1^-(Z) \leqslant E(f^{1/2}) \leqslant E(1_A f^{1/2})$$

$$\cdot \leqslant \{P(A)E(f)\}^{1/2} \leqslant (1/ab)\{P(A)\bar{b}^2 P(A)\}^{1/2}$$

$$\mathbf{N}_1^-(Z) \leqslant 1 + \epsilon.$$

On the other hand, using the fact that $E(X_1(X_1 - X_u)1_A) = E((X_1 - X_u)^2 1_A)$,

$$b = (1/ab)b^2 P(A) \leqslant \langle X, Z\rangle.$$

This inequality and (3) *imply*

$$\mathbf{N}_\infty(X) \leqslant \sqrt{5} \sup_{Z\in K} \langle X, Z\rangle.$$

(5) Let f be an element of $(\mathbf{H}_1)'$ the topological dual of $\mathbf{H}_1$. Then f induces a continuous linear mapping on $\mathfrak{M}^2$. Thus there exists an element X of $\mathfrak{M}^2$ such that $\langle X, Z\rangle = f(Z)$ for all the elements Z of $\mathfrak{M}^2$. This implies

$$\sup_{Z\in K} \langle X, Z\rangle \leqslant \|f\|_{(\mathbf{H}_1)'} < +\infty$$

and

$$\mathbf{N}_\infty(X) \leqslant \sqrt{5}\, \|f\|_{(\mathbf{H}_1)'}$$

according to (4) above.

Then, we have defined a mapping $f \rightsquigarrow X$ from $(\mathbf{H}_1)'$ into BMO. Now X induces an element $\bar{f}$ of $(\mathbf{H}_1)'$ by the formula

$$\bar{f}(Z) = \langle X, Z\rangle$$

(see (1)) and we have $f(Z) = \bar{f}(Z)$ for every element Z of $\mathfrak{M}^2$; this implies that $f(Z) = \bar{f}(Z)$ for all elements Z of $\mathbf{H}_1$ (see (2)) and thus completes the proof.

INEQUALITIES FOR SEMIMARTINGALES

11.9 Lemma

Lemma *Let $(V_t)_{t\in T}$ be a real Meyer process V, i.e. (see Section 9), a real cadlag predictable process of bounded variation such that $E\{\int_T d|V_t|\} < +\infty$. Then there exist two disjoint predictable sets A and B such that if we define $Y = 1_A - 1_B$, we have*

$$E\left\{\int_T d|V_t|\right\} = E\left\{\int_T Y\,dV_t\right\}.$$

Moreover, $\int_T Y\,dV_t$ is a positive random variable.

Proof Let v be the Doléans measure associated with V. Since this is a real measure, there exist two disjoint predictable sets A and B and two positive measures v^+ and v^- such that $v = v^+ - v^-$, and the support of v^+ (resp. v^-) is A (resp. B). If we define $Y = 1_A - 1_B$, we have

$$E\left\{\int_T d|V_t|\right\} = v^+(A) + v^-(B) = \int Y\,dv = E\left\{\int_T Y_t\,dV_t\right\}.$$

11.10 Proposition

Proposition *Let S be a real cadlag semimartingale, i.e., $S = V + M$, where V is an adapted cadlag process of bounded variation and M is a cadlag local martingale. We define*

$$\mathbf{N}_1''(S) := E\left\{\int_T d|V_t|\right\} + \mathbf{N}_1'(M)$$

if $S = V + M$, where V is a (unique and real) Meyer process and M a martingale (of course, $\mathbf{N}_1'(M)$ is defined as in Section 11.1),

$$\mathbf{N}_1''(S) := +\infty \textit{ if such a decomposition does not exist;}$$

$$\mathbf{N}_1'(S) := \sup E\left(\int Y\,dS\right),$$

this supremum being taken for all the real predictable processes Y uniformly bounded by 1. Then $\mathbf{N}_1''$ and $\mathbf{N}_1'$ are two equivalent seminorms. More precisely, for every real semimartingale S we have

$$\mathbf{N}_1'(S) \leqslant \mathbf{N}_1''(S) \leqslant 3\mathbf{N}_1'(S).$$

Proof (1) It is easily seen that $\mathbf{N}_1''$ and $\mathbf{N}_1'$ are seminorms. Moreover, if M is a martingale, $\mathbf{N}_1''(M) = \mathbf{N}_1'(M)$ (because $M = V + M'$, where M' is a martingale and V a Meyer process implies that $V = 0$).

We also have $\mathbf{N}_1''(V) = \mathbf{N}_1'(V)$ if V is a Meyer process (see Section 11.9).

(2) Let S be a real martingale such that $\mathbf{N}_1''(S) < +\infty$. Then $S = V + M$, where V is a Meyer process and M is a martingale. Moreover,

$$\mathbf{N}_1''(S) = \mathbf{N}_1'(V) + \mathbf{N}_1'(M) \qquad \text{(see (1) above).}$$

But there exists a predictable process Y uniformly bounded by 1 such that

$$\mathbf{N}_1'(V) = E\left[\int_T Y\,dV\right]$$

and $\int_T Y\,dV$ is a positive random variable. Then

$$E\left(\left|\int_T Y(dV + dM)\right|\right) \geqslant E\left(\int_T Y\,dV\right)$$

(because $E(\int Y\,dM) = 0$), i.e., $\mathbf{N}_1'(V + M) \geqslant \mathbf{N}_1'(V)$.

We also have

$$\mathbf{N}_1'(V + M) + \mathbf{N}_1'(V) \geqslant \mathbf{N}_1'(M).$$

. This implies that

$$3\mathbf{N}_1'(S) = 3\mathbf{N}_1'(V + M) \geqslant \mathbf{N}_1'(M) + \mathbf{N}_1'(V) = \mathbf{N}_1''(S).$$

(3) Let S be a real semimartingale such that $\mathbf{N}_1'(S) < +\infty$. This implies that the Doléans function of S is a Doléans measure, and $S = V + M$, where V is a Meyer process and M is a martingale. Moreover,

$$\begin{aligned}\mathbf{N}_1'(S) &\leqslant \mathbf{N}_1'(V) + \mathbf{N}_1'(M)\\ &\leqslant E\left\{\int_T d|V_t|\right\} + \mathbf{N}_1'(M) \qquad \text{(see Section 11.9)}\\ &\leqslant \mathbf{N}_1''(S).\end{aligned}$$

HISTORICAL NOTES

Section 8. The inequalities and equi-integrability properties contained in this section are due essentially to Doob (cf. [Doo-2]). The existence of the "Doléans-measure" for submartingales of class D was first proved in [Dol-4]. The following details may be found in [MeP-1]: equivalence of this existence and being of class D (when the filtration $\mathcal{F}_t$) is complete); the weaker and equivalent condition for being of class D (Corollary 8.5); the formulation in Section 8.7 of the stopping theorem. For different conditions of existence of the Doléans measure when the filtration $(\mathcal{F}_t)$ is not complete the reader can consult [Föl].

Section 9. The initial Meyer decomposition theorems for supermartingales (see [Mey-6] and [Mey-1]) were considerably clarified and simplified by the use of Dellacheries' section and projection theorems (see [Del-1]). This approach easily extends to Hilbert-valued processes (see an exposition in [Met-6]). In this book, we took a kind of short cut, avoiding the whole theory of projection and section and restricting ourselves to using a particular case of the projection theorem under the form of the directly proved lemma of Section 9.8.

The Meyer process of a Banach-valued admissible measure was built in [Pel-2]. The presentation given here is much simplified.

Section 10. The starting results on the structure of square integrable martingales can be found in [Mey-4, Chapter II]. The main result of this section, concerning the relation between π^*-processes and semimartingales, has been proved by the authors in [McP-5] and [MeP-6].

Prelocally p-summable processes, which are the same as semimartingales when the processes are real, were introduced by Kussmaul [Kus].

Section 11. The problem of the equivalence of norms for Banach spaces of martingales has been a lively topic for some years. The first important inequalities related to H^p spaces ($p > 1$) were obtained by Burkholder [Bur]. Several authors (Burkholder, Davis, and Gundy) gave simplifications and extensions. Davis, in particular, considered the most difficult and interesting case, $p = 1$. Garsia remarked that the Fefferman inequality could be used to deduce the Davis inequality. A systematic exposition of those inequalities in the case of continuous time can be found in [Mey-4 Chapter IV]. We have reconsidered here the exposition of the whole set of inequalities (only in the case $p = 1$), trying to give proofs that are as direct as possible and staying, by the way, pretty close to the initial idea of Burkholder (see, for example, the new proof of the inequality of Section 11.5 in the text).

The study of the duality between $\mathbf{H}^1$ and BMO, discovered by Fefferman, follows essentially the Meyer presentation.

Inequalities for semimartingales originate in [Yor-3].

CHAPTER 5

STOCHASTIC MEASURES

In this chapter we study stochastic integration from a very natural point of view, which we have often considered. The only reason to defer this until now was our concern with taking a starting point as elementary as possible and dealing as soon as possible with stochastic equations.

The integral of elementary processes $Y \in \mathcal{E}$, as defined in Section 2.2, can indeed be clearly interpreted as an integral, in the classical sense, of Y with respect to an additive function taking its values in the topological vector space $L^0_{\mathbf{B}}(\Omega, \mathcal{F}, P)$.

Then it is possible to use all the classical results on vector measures and integrals. More precisely, the stochastic integral $\int Y\,dX$ can be defined as the integral of the real function Y on $(\Omega \times T)$ with respect to the additive function x defined on $\mathcal{Q}$ (see Section 1.7) by $x(A) := \int 1_A\,dX$.

Stochastic measures are thus measures with values in $L^0_{\mathbf{B}}(\Omega, \mathcal{F}, P)$, possessing some additional property which expresses the stochastic flavor of the concept (see Section 12.2 below).

Section 12 is devoted to the extension theorems and σ-additivity properties of stochastic measures (Daniell theorem, dominated convergence theorem, etc.). It ends up with an important, very recently proved theorem by Dellacherie–Mokobodski–Meyer, which essentially states that L^0-stochastic measures are associated with semi-martingales, when $\mathbf{B} = \mathbf{R}$ (and T is an interval of $\mathbf{R}$: see below).

In section 13, a Riesz representation theorem is proved for stochastic measures. It essentially states that a functional definition of stochastic measures as "Radon" stochastic measures can be given.

An interesting feature of the development in this chapter is that in our general setting we can consider an index set T, which, instead of always being a subset of $\overline{\mathbf{R}}_+$, may be a quite general topological space.

12 STOCHASTIC MEASURES AND RELATED INTEGRATION

12.1 General Hypotheses

Throughout Chapter 5 **B** is a Banach space and $\mathcal{B}$ its Borel σ-algebra, $(\Omega, \mathcal{F}, P)$ is a complete probabilized space, T is a topological space, and $\mathcal{J}$ is a boolean semialgebra of subsets of T (i.e., $T \in \mathcal{J}$ and for every J and J' in $\mathcal{G}$, the set $J \cap J'$ belongs to $\mathcal{G}$ and $J - J \cap J'$ is a finite union of disjoint elements of $\mathcal{J}$). For every element J of $\mathcal{J}$, $\mathcal{F}_J$ is a sub-σ-algebra of $\mathcal{F}$ which includes all the P-null elements of $\mathcal{F}$. Moreover, we assume that $J \subset J'$ implies $\mathcal{F}_{J'} \subset \mathcal{F}_J$.

When $T = [0, 1]$ is the unit interval of the real line and $(\Omega, \mathcal{F}, P(\mathcal{F}_t)_{t \in T})$ is a complete stochastic basis (see Section 1.1), the previous assumptions are fulfilled when we denote by $\mathcal{J}$ the semialgebra of subsets of T generated by the intervals $]0, t]$, $t \in T$, and when we put for every element J of $\mathcal{J}$, $\mathcal{F}_J = \mathcal{F}_t$ if $t = \inf\{s : s \in J\}$. For other examples see [Met-5].

We denote by $\mathcal{R}$ the family of the subsets of $(\Omega \times T)$ so-called "predictable rectangles," i.e., A belongs to $\mathcal{R}$ if and only if $A = (F \times J)$, where J is an element of $\mathcal{J}$ and F belongs to $\mathcal{F}_J$.

It is easily seen that $\mathcal{R}$ is a semialgebra. We denote by $\mathcal{A}$ the algebra generated by this semialgebra; it is easily seen, as in Section 1.8, that every element A of $\mathcal{A}$ admits a partition constituted of elements of $\mathcal{R}$.

According to the previous example, when $T = [0, 1]$, these notations, in the general setting considered here, are consistent with those introduced in Section 1.

The σ-algebra generated by $\mathcal{R}$, or $\mathcal{A}$, will be denoted by $\mathcal{P}$ and called the σ-algebra of predictable sets. A real function defined on $(\Omega \times T)$ will be called a predictable process if it is $\mathcal{P}$-measurable. The vector space of all the real $\mathcal{A}$-simple processes will be denoted by $\mathcal{E}$; moreover, we put

$$\mathcal{E}_0 := \left\{ Y : Y \in \mathcal{E}, \sup_{\omega, t} |Y_t(\omega)| \leqslant 1 \right\}.$$

Let x be an additive $L^p_{\mathbf{B}}$-valued function defined on $\mathcal{A}$.

When Y belongs to $\mathcal{E}$, namely, $Y := \sum_{i \in I} a_i 1_{A(i)}$, we define the stochastic integral $\int Y\,dx$ in the usual way, i.e.,

$$\int Y\,dx := \sum_{i \in I} a_i x(A(i)).$$

For each element p of $[0, \infty]$, we put $L^p_{\mathbf{B}} := L^p_{\mathbf{B}}(\Omega, \mathcal{F}, P)$ (complete metric space, or Banach space if $p \geqslant 1$, endowed with its usual topology).

A subset $\mathcal{U}$ of $L^p_{\mathbf{B}}$ will be called a bounded subset of $L^p_{\mathbf{B}}$ if this property holds in the usual sense for topological vector spaces (see [Bou-2] or [Tre]).

12.2 Stochastic Measures

Definition Let $p \geqslant 0$. Let x be an additive $L^p_{\mathbf{B}}$-valued function defined on $\mathcal{Q}$ such that for every element $(F \times J)$ of $\mathcal{R}$, $x(F \times J) = 1_F x(F \times J)$, and such that the set $x(\mathcal{Q}) := \{x(A), A \in \mathcal{Q}\}$ is bounded. Such an additive function x will be called an *$L^p_{\mathbf{B}}$-stochastic measure* if there is an extension, necessarily unique, of x to the σ-algebra $\mathcal{P}$ which is σ-additive for the usual (strong) toplogy of $L^p_{\mathbf{B}}$. In this case this extension will still be denoted by x; moreover, if F and A belong to $\mathcal{F}$ and $\mathcal{P}$, respectively, with $A \subset (F \times T)$, we have clearly $x(A) = 1_F x(A)$ (indeed, this property is fulfilled for every element A of $\mathcal{R}$).

As already noted, the purpose of this section is to give necessary and sufficient conditions for x to be an $L^p_{\mathbf{B}}$-stochastic measure (see the extension theorem of Section 12.7 below) and to construct the stochastic integral $(\int Y\,dX)$ of a real process Y with respect to such a stochastic measure x.

To prove such a theorem, some preliminary lemmas are needed.

12.3 σ-Continuity Lemma for a Subadditive Measure

Lemma *Let v be a nonnegative function defined on $\mathcal{Q}$ fulfilling the following three properties*:

(i) *for all the elements A and A' of $\mathcal{Q}$, $v(A) \leqslant v(A \cup A') \leqslant v(A) + v(A')$*;

(ii) *for $\epsilon > 0$ and for every element $A = F \times J$ of $\mathcal{R}$ there exists an element J' of $\mathcal{J}$ and a compact subset K of T such that $J' \subset K \subset J$ and, if $A' := F \times J'$, $v(A \backslash A') \leqslant \epsilon$*;

(iii) *for every decreasing sequence $(F_n)_{n>0}$ of elements of $\mathcal{F}$ with $F_n \downarrow \varnothing$ and for every associated sequence $(A_n)_{n>0}$ of elements of $\mathcal{Q}$ such that for all the integers n, $A_n \subset (F_n \times T)$, we have $\lim_{n\to\infty} v(A_n) = 0$.*

Then the following property holds:

(iv) *for every decreasing sequence $(A_n)_{n>0}$ of elements of $\mathcal{Q}$ with $A_n \downarrow \varnothing$ we have $\lim_{n\to\infty} v(A_n) = 0$.*

Proof The proof (see [MeP-3]) is quite analogous to the proof of the lemma of Section 8.4: it is left to the reader.

12.4 The F-Norm $\|\cdot\|_p$

Let $p \geqslant 0$. For every random variable f belonging to $L^p_{\mathbf{B}}$, we put

$$\|f\|_p := \left[\int |f(\omega)|^p P(d\omega)\right]^{1/p} \quad \text{if} \quad p \geqslant 1 \qquad \text{(usual Banach norm)},$$

$$\|f\|_p := \int |f(\omega)|^p P(d\omega) \quad \text{if} \quad 0 < p \leqslant 1,$$

$$\|f\|_0 := \int (|f(\omega)| \wedge 1) P(d\omega).$$

Then $\|\cdot\|_p$ is an F-norm associated with the usual toplogy of $L^p_{\mathbf{B}}$: $\|f\|_p = 0$ if $f = 0$, $\|f\|_p = \|-f\|_p$, $\|f+g\|_p \leqslant \|f\|_p + \|g\|_p$ and $\lim_n \|\lambda_n f_n - \lambda f\|_p = 0$ for every sequence of scalars λ_n and of $f_n \in L^p_{\mathbf{B}}$ such that $\lim_n \|f_n - f\|_p = 0$ and $\lim_n |\lambda_n - \lambda| = 0$.

12.5 A Bounded Additive Function

Lemma *Let $\|\cdot\|_p$ be the F-norm defined on $L^p_{\mathbf{B}}$ as above and x be an $L^p_{\mathbf{B}}$-valued additive function defined on an algebra $\mathcal{A}$. Let v be the function defined on $\mathcal{A}$ by*

$$v(A) := \sup_{B \in \mathcal{A},\, B \subset A} \|x(B)\|_p.$$

We consider the following three properties for x:

(i) *for every sequence $(A_n)_{n>0}$ of pairwise disjoint elements of $\mathcal{A}$, $\lim_{n\to\infty} x(A_n) = 0$,*

(ii) *for every element A of $\mathcal{A}$, $v(A) < +\infty$;*

(iii) *$x(\mathcal{A})$ is a bounded subset of $L^p_{\mathbf{B}}$.*

Then (i) *implies* (ii) *and, when $p > 0$,* (ii) *implies* (iii).

Proof That assertion (ii) implies (iii) is obvious; that (i) implies (ii) is left for the reader to prove (see Corollary 4.11 in [Dre]).

12.6 Daniell-Type Theorem

In this theorem Ω is a set, $\mathbf{B}$ is a Banach space, p is a nonnegative real number, and $\mathcal{G}$ is a vector space of real functions on Ω such that if g and g' belong to $\mathcal{G}$, $(g \wedge g')$ belongs to $\mathcal{G}$.

Theorem *Let m be an $L^p_{\mathbf{B}}$-valued linear mapping defined on $\mathcal{G}$ for which the following two properties are fulfilled:*

(i) *for every decreasing sequence $(g_n)_{n>0}$ of elements of $\mathcal{G}$ such that $g_n \downarrow 0$ the sequence $m(g_n)$ converges to zero (for the usual topology of $L^p_{\mathbf{B}}$);*

(ii) *for every sequence* $(g_n)_{n \geqslant 0}$ *of nonnegative elements of* $\mathcal{G}$ *such that* $\sum_{n>0} g_n \leqslant g_0$ *the sequence* $m(g_n)$ *converges to zero (for the usual topology of* $L^p_{\mathbf{B}}$*).*

Then the mapping m *can be extended into an* $L^p_{\mathbf{B}}$*-valued linear mapping for which the Lebesgue dominated convergence theorem holds. In particular, if* $\mathcal{G} := \mathcal{E}$ *(as defined in Section* 12.1*),* m *can be extended to the vector space of all uniformly bounded predictable processes.*

Proof This theorem is proved in [Pel-1]. It is a purely measure-theoretic result, which we admit here. (See also [Sio].)

12.7 Extension and Dominated Convergence Theorem

Theorem *We consider the hypotheses and notations given in Section* 12.1. *Let* $p \geqslant 0$ *and let* x *be an* $L^p_{\mathbf{B}}$*-valued additive function defined on* $\mathcal{Q}$. *For every subset* A *of* $(\Omega \times T)$, *we put*

$$v(A) := \sup_{B \subset A,\, B \in \mathcal{Q}} \|x(B)\|_p \qquad (\textit{see Section } 12.4).$$

We assume that x *satisfies the following four properties:*

(i) *if* $A := (F \times J)$ *belongs to* $\mathcal{R}$, $x(A) = 1_F x(A)$;

(ii) *for every* $\epsilon > 0$ *and every element* $A := F \times J$ *of* $\mathcal{R}$ *there exists an element* J' *of* $\mathcal{J}$ *and a compact subset* K *of* T *such that* $J' \subset K \subset J$ *and* $v(A \backslash A') \leqslant \epsilon$ *with* $A' = F \times J'$;

(iii) *for every sequence* $(A_n)_{n>0}$ *of pairwise disjoint elements of* $\mathcal{Q}$ *the sequence* $x(A_n)$ *goes to zero in* $L^p_{\mathbf{B}}$;

(iii′) $x(\mathcal{Q})$ *is a bounded subset of* $L^p_{\mathbf{B}}$.

Then x *is an* $L^p_{\mathbf{B}}$*-stochastic measure. Moreover, the mapping* $Y \mapsto (\int Y\,dX)$ *defined on* $\mathcal{E}$ *in the usual way (see Section* 12.1*) can be extended into a (unique)* $L^p_{\mathbf{B}}$*-valued linear mapping, defined on the set of all uniformly bounded predictable processes* Y *such that the following dominated convergence property holds:*

(iv) *if* $(Y)_{n>0}$ *is a uniformly bounded sequence of real predictable processes, which converges to* Y, *we have*

$$\int Y\,dx = \lim_{n\to\infty} \int Y_n\,dx.$$

If B *is a finite-dimensional vector space, property* (iii) *holds as soon as* x *is an additive function and* $x(\mathcal{Q}) := \{x(A), A \in \mathcal{Q}\}$ *is a bounded subset of* $L^p_{\mathbf{B}}$.

Proof In the sequel of this proof, we write $\|\cdot\|$ instead of $\|\cdot\|_p$.

(1) At first we consider the case $\mathbf{B} = \mathbf{R}$.

We recall the following properties for spaces $L^p_{\mathbf{R}}$ (for a proof see [MaO] for example): for every sequence (x_n) in $L^p_{\mathbf{R}}$, which is "perfectly bounded" (i.e., such that the family $\{\sum_{x_n \in A} \epsilon_n x_n : A \text{ finite } \epsilon_n = \pm 1\}$ is bounded in $L^p_{\mathbf{R}}$), the series $\sum_n x_n$ converges in $L^p_{\mathbf{R}}$. Let $(A(n))_{n>0}$ be a sequence of disjoint elements of $\mathcal{C}$: since x is an additive function and $x(\mathcal{C})$ is a bounded subset of $L^p_{\mathbf{B}}$, the series $\sum_n (x(A(n))$ is "perfectly bounded"; thus it is convergent. This proves the end of the theorem when $\mathbf{B} = \mathbf{R}$, and we have the same property when $\mathbf{B}$ is a finite-dimensional vector space.

(2) Now we suppose that $\mathbf{B}$ is a general Banach space. First we prove that the restriction of v to $\mathcal{C}$ satisfies the properties (i), (ii), and (iii) of the lemma of Section 12.3. Condition (i) of Section 12.3 is obviously satisfied and condition (ii) from Section 12.3 is condition (ii) of Section 12.7. If condition (iii) of Section 12.3 is not fulfilled, there exist $\epsilon > 0$, a decreasing sequence $(F(n))_{n>0}$ of elements of $\mathcal{F}$ with $F(n)\downarrow\emptyset$, and a sequence $(A_n)_{n>0}$ of elements of $\mathcal{C}$ such that for each integer n, $A_n \subset (F(n) \times T)$ and $\|x(A_n)\| \geqslant 8\epsilon$. We have $v(F(1) \times T) = a < +\infty$ (cf. Section 12.5). Let D be a set such that $D \subset (F(1) \times T)$, $D \in \mathcal{C}$, and $\|x(D)\| \geqslant a - \epsilon$; let k be an integer such that $k > 1$ and $\|x(D)1_{F(k)}\| \leqslant \epsilon$; let E be a set that belongs to $\mathcal{C}$ and such that $E \subset (F(k) \times T)$ and $\|x(E)\| \geqslant 8\epsilon$; we have

$$\|X(E \backslash D)\| \geqslant 4\epsilon \quad \text{or} \quad \|x(E \cap D)\| \geqslant 4\epsilon.$$

In the first case we have

$$\|x(E \cup D)\| \geqslant a + \epsilon - \epsilon + (4\epsilon - \epsilon) \geqslant a + 3\epsilon,$$

and in the second one we have

$$\|x(D \backslash E)\| \geqslant a + 2\epsilon.$$

In the two cases, this implies $v(F(1) \times T) \geqslant a + 2\epsilon$; but this is impossible; then condition (iii) of Section 12.3 is fulfilled, and we can apply the lemma in Section 12.3.

(3) Now we prove property (i) of Section 12.6.

First, the set $\{Z : Z = \int Y\,dx,\ Y \in \mathcal{E}_0\}$ is a bounded subset of $L^p_{\mathbf{B}}$. This follows easily from (iii), (iii′), and the properties of the F-norm considered here for $p \geqslant 1$. For $p < 1$ it is slightly more subtle—see [Tur]. That means that for each $\epsilon > 0$, there exists $\eta(\epsilon)$ such that $\sup_{t,\omega} |Y_t(\omega)| \leqslant \eta(\epsilon)$ implies $\|\int Y\,dX\| \leqslant \epsilon$. We fix ϵ and $\eta(\epsilon)$.

Let $(Y_n)_{n>0}$ be a sequence of nonnegative $\mathcal{C}$-simple processes such that $Y_n\downarrow 0$. For every integer n we put $A(n) := [Y_n > \eta(\epsilon)]$ and $B(n) = \Omega \backslash A(n)$. We have $\|\int Y_n 1_{B(n)}\,dx\| \leqslant \epsilon$ (because $Y_n 1_{B(n)} \leqslant \eta(\epsilon)$). Moreover, $A(n)\downarrow\emptyset$; thus $v(A(n))\downarrow 0$ (cf. (2) above). Then $\lim_{n\to\infty} \int Y_n 1_{A(n)}\,dx = 0$, and that proves condition (i) of Section 12.6.

(4) Now we prove that condition (ii) of Section 12.6 is fulfilled. Let $(Y_n)_{n>0}$ be a sequence of nonnegative $\mathcal{C}$-simple processes such that

$\sum_{n>0} Y_n \leqslant 1$. Let ϵ be a positive real number. We consider $\eta(\epsilon)$ as in (3) above, and we can suppose that $\eta(\epsilon) = 1/k$ (k depends on ϵ). For every integer n we put $A(n) := [Y_n > \eta(\epsilon)]$. As in (3) above, we have to prove that

$$\lim_{n\to\infty} \int Y_n 1_{A(n)}\, dx = 0.$$

For that we consider the sequences of sets

$$B(0,0) = \Omega, \qquad B(n,-1) = \emptyset \quad \text{if} \quad n \geqslant 0,$$

$$B(0,j) = \emptyset \quad \text{if} \quad j \geqslant 1$$

and for $n \geqslant 1$ and $j \geqslant 1$

$$B(n,j) := [B(n-1,j-1) \cap A(n)] \cup [B(n-1,j) \backslash A(n)],$$

$$C(n,j) := B(n-1,j-1) \cap A(n) = B(n,j) \cap A(n).$$

If $j > k$, we have $B(n,j) = \emptyset$; moreover, for every integer n $\{C(n,j)\}_{1 \leqslant j \leqslant k}$ is a partition of $A(n)$, and for every integer j the sets $\{C(n,j)\}_{n>0}$ are disjoint. It follows from (iii) that

$$\lim_{n\to\infty} \int Y_n 1_{C(n,j)}\, dx = 0$$

for every integer j with $1 \leqslant j \leqslant k$. But this implies $\lim_{n\to\infty} \int Y_n 1_{A(n)}\, dX = 0$ (because $\{C(n,j)\}_{1 \leqslant j \leqslant k}$ is a partition of $A(n)$), and that proves (ii) of Section 12.6.

Then we can apply the Daniell theorem of Section 12.6, and that completes the proof.

12.8 Proposition

Proposition *We consider the hypotheses of Section* 12.7; *with the same notation, assertion* (ii) *of Section* 12.7 *holds when* (iii) *of Section* 12.7 *and the following two properties are verified*:

(v) *for every element J of $\mathcal{J}$ there exists a sequence $(J_n)_{n>0}$ of elements of $\mathcal{J}$ and an associated sequence $(K_n)_{n>0}$ of compact subsets of T such that $J_n \uparrow J$ and for all integers n, $J_n \subset K_n \subset J$;*

(vi) *for every sequence $(A_n)_{n>0}$ of elements of $\mathcal{R}$, with $A_n \downarrow \emptyset$, $\lim_{n\to\infty} x(A_n) = 0$ (for the usual topology of $L^p_{\mathbf{B}}$).*

Proof Let us assume that there exist $\epsilon > 0$ and an element $A := F \times J$ of $\mathcal{R}$ such that (ii) of Section 12.7 does not hold.

Let $(J_n)_{n>0}$ be a sequence of elements of $\mathcal{J}$ and $(K_n)_{n>0}$ be an associated sequence of compact subsets of T such that for all integers n, $J_n \subset K_n \subset J$ and $J_n \uparrow J$. For every integer n we put $J_n^* := J \backslash J_n$.

Now we define recurrently the sequence $(V_n, W_n)_{n>0}$ of pairs of elements of $\mathcal{Q}$ and the associated sequence $(k(n))_{n>0}$ of integers as follows:

$$k(0) := 1, \qquad W_{n+1} \subset (F \times J^*_{k(n)}), \qquad \|x(W_{n+1})\|_p \geqslant \tfrac{1}{2}\epsilon;$$

$k(n+1)$ is chosen such that

$$\|x[W_{n+1} \cap (F \times J^*_{k(n+1)})]\|_p \leqslant \tfrac{1}{4}\epsilon$$

(such an integer $k(n+1)$ exists according to condition (vi) and because there exists a finite partition of W_{n+1} that is constituted of elements of $\mathcal{Q}$).

We put

$$V_{n+1} := [W_{n+1} \backslash (F \times J^*_{k(n+1)})].$$

We have

$$\begin{aligned}\|x(V_{n+1})\|_p &\geqslant \|x(W_{n+1})\|_p - \|x[W_{n+1} \cap (F \times J^*_{k(n+1)})]\|_p \\ &\geqslant \tfrac{1}{2}\epsilon - \tfrac{1}{4}\epsilon = \tfrac{1}{4}\epsilon.\end{aligned}$$

But the sets $(V_n)_{n>0}$ are pairwise disjoint, which contradicts property (iii) of Section 12.7 and completes the proof.

12.9 Remark

Remark We consider the hypotheses and notation of the theorem of Section 12.7; let us assume that all the conditions of this theorem are fulfilled. Let Y be a uniformly bounded predictable process. Then the mapping z defined by $z(A) = \int 1_A Y\,dx$ for every predictable set A is an $L^p_{\mathbf{B}}$-stochastic measure.

12.10 Minimax Lemma

Lemma *Let $C > 0$ and G be a convex part of L^1. Let K be a convex compact set for the topology $\sigma(L^\infty, L^1)$ such that for every element g of G there exists an element k of K with $\langle g, k\rangle \leqslant C$. Then there exists an element h of K such that for every element g of G, $\langle h, g\rangle \leqslant C$.*

Proof Actually, this lemma is a particular case of a general topological lemma where L^∞ and L^1 may be any locally convex spaces in duality. According to the compactness of K, it is sufficient to consider the case in which G is generated by a finite family $(g_i)_{i\in I}$ (for all such families consider the closed subset $\overline{C}$ of K defined by $\overline{C} := \{k : \langle k, g_i\rangle \leqslant C, \forall i \in I\}$).

Let us assume that there exists $\epsilon > 0$ such that for every element k of K there exists $i \in I$ with $\langle k, g_i\rangle \geqslant C + \epsilon$. We derive a contradiction.

In other words, in the space $\mathbf{R}^I$, $A \cap B = \emptyset$, where

$$A := \{ y : y \in R^I, \exists k \in K, \forall i \in I, y_i = \langle k, g_i \rangle \}$$

and

$$B := \{ y : y \in R^I, \forall i \in I, y_i \leqslant C \}.$$

A (resp. B) is a compact (resp. closed) convex subset of $\mathbf{R}^I$. Then there exists a continuous linear mapping x on $\mathbf{R}^I$ such that

$$\sup_{y \in B} x(y) < \inf_{y \in A} x(y).$$

If $x(y) = \sum_{i \in I} x_i y_i$, we have $x_i \geqslant 0$ for every element i of I; now we put $a = \sum_{i \in I} x_i$ and $g = (1/a) \sum_{i \in I} x_i g_i$. According to the convexity of G, g belongs to G and there exists $k \in K$ such that $\langle g, k \rangle \leqslant C$ and $\inf_{y \in A} x(y) \leqslant Ca$. Since clearly, $\sup_{x \in B} x(y) = Ca$, we have a contradiction and this completes the proof.

12.11 Fundamental Property of Bounded Sets in L_0 [1]

Theorem *Let $(\Omega, \mathfrak{F}, P)$ be a complete probabilized space. Let G be a convex part of $L^1(\Omega, \mathfrak{F}, P)$ that is bounded in $L^0(\Omega, \mathfrak{F}, P)$. Then there exists an element f of $L^1(\Omega, \mathfrak{F}, P)$ such that $P[f > 0] = 1$ and*

$$\sup_{g \in G} \int_\Omega g(\omega) f(\omega) dP(\omega) \leqslant 1.$$

Proof For every integer n we define

$$K^n := \left\{ k : k \in L^\infty, 0 \leqslant k \leqslant 1, \int_\Omega k(\omega) P(d\omega) \geqslant 1 - (1/n) \right\},$$

where K^n is a convex compact set for the topology $\sigma(L^\infty, L^1)$; moreover, since G is bounded in L^0, there exists $C(n) > 0$ such that for every element g of G there exists an element k of K^n with $\int_\Omega kg\, dP \leqslant C(n)$ (choose $C(n)$ such that $P[|g| > C(n)] \leqslant 1/n$ and $k = 1 - 1_{|g| > C(n)}$). Now according to the previous lemma there exists an element h_n of K^n such that for every element g of G

$$\langle h_n, g \rangle = \int_\Omega g h_n\, dP \leqslant C(n).$$

Then the theorem is proved by setting

$$f := \sum_{n > 0} 2^{-n} \frac{1}{C(n)} h_n.$$

[1] See [Nik] and [Mau].

12.12 Dellacherie–Meyer–Mokobodski Theorem

Theorem *Let* **B** *be a finite-dimensional vector space and* X *a* **B**-*valued cadlag adapted process* (*see the theorem of Section* 1.16) *indexed by* $T \subset \mathbf{R}_+$. *Then the following two properties are equivalent*:

(i) X *is a semimartingale* (*see Section* 10.8);
(ii) *the set* $G := \{Z : Z = \int 1_A\, dX,\ A \in \mathcal{Q}\}$ *is bounded in* $L^0(\Omega, \mathfrak{F}, P)$.

Proof Let us recall that X is a real semimartingale if and only if X is a π-process, or a π^*-process or prelocally 1-summable (see Section 10.9). Thus if X is a semimartingale, the set G is clearly bounded in L^0.

Let X be a cadlag adapted process such that G is bounded. We have only to consider the case where X is real valued. Let V be the finite (X is cadlag) random variable defined by $V := \sup_{t \in T}|X_t|$. For every element A of $\mathcal{Q}$ and every stopping time u

$$\left|\int 1_A\, dX^u\right| \leqslant 2V + \left|\int 1_A 1_{]0,\,u]}\, dX\right|,$$

where X^u is the process X stopped "strictly before" the stopping time u (see Section 1.13). Thus by prelocalization we can suppose that X is uniformly bounded.

In this case we put $G := \{Z : Z = \int Y\, dX,\ Y \in \mathcal{E}_0\}$; this set is bounded in $L^0(\Omega, \mathfrak{F}, P)$ and contained in $L^1(\Omega, \mathfrak{F}, P)$, and we can apply the theorem of Section 12.11. Then there exists a probability Q equivalent to P such that for this probability Q the Doléans function $d(X)$ is bounded.

According to the boundedness of X, it is easy to verify (see Section 12.3) that this function $d(X)$ is σ-additive: this step is left to the reader.

Then $X = A + M$, where A is the Meyer process associated with $d(X)$ and M is a Q-martingale. Thus M is a π-process for P (see Section 4.8) and X is a π-process for P or, equivalently, a semimartingale.

13 RIESZ REPRESENTATION THEOREM

STOCHASTIC LINEAR FUNCTIONAL

13.1 Hypothesis and Notations

Throughout Section 13 we assume that all the hypotheses given in Section 12.1 are fulfilled. Moreover, $\mathcal{G}$ is a vector space of uniformly bounded real predictable processes such that $\mathcal{P}$ is generated by $\mathcal{G}$, $1_{(\Omega \times T)}$ belongs to $\mathcal{G}$, and if g_1 and g_2 belong to $\mathcal{G}$, then $g_1 \wedge g_2$ belongs to $\mathcal{G}$. This space $\mathcal{G}$ is assumed to be a Banach space for the uniform norm which will be denoted by $\|\cdot\|$, i.e.,

$$\|g\| = \sup_{t,\,\omega} |g(\omega, t)|.$$

For example, $\mathcal{G}$ can be the space of all real predictable processes such that for every element ω of Ω the function $t \mapsto g(\omega, t)$ is uniformly continuous. For other examples see [Met-5].

13.2 Stochastic Linear Functional

Definition An $L^p_{\mathbf{B}}$-valued function m defined on $\mathcal{G}$ will be called an $L^p_{\mathbf{B}}$-stochastic linear function (with $p \geqslant 0$) if m is an $L^p_{\mathbf{B}}$-valued linear continuous function defined on $\mathcal{G}$, $\mathcal{G}$ being endowed with its uniform topology and $L^p_{\mathbf{B}}$ with its usual (strong) topology, which fulfills the following property:

(i) if g belongs to $\mathcal{G}$, F belongs to $\mathcal{F}$, and $1_F g$ belongs to $\mathcal{G}$, $m(1_F g) = 1_F m(1_F g)$.

The relation between such stochastic linear functionals and stochastic measures is established in the following theorem.

13.3 Representation Theorem

Theorem (1) *Let $p \geqslant 0$. Let x be an $L^p_{\mathbf{B}}$-stochastic measure (then (see Section 12.5) $x(\mathcal{Q})$ is a bounded subset of $L^p_{\mathbf{B}}$). We denote by m the linear mapping on $\mathcal{G}$ defined by $m(g) = \int g\,dX$. Then m is an $L^p_{\mathbf{B}}$-stochastic linear functional.*

(2) *Now let m be an $L^p_{\mathbf{B}}$-stochastic linear function for which the following statement holds:*

(i) *for every sequence $(g_n)_{n>0}$ of nonnegative elements of $\mathcal{G}$ such that $\sum_{n>0} g_n \leqslant 1$ we have $\lim_{n\to\infty} m(g_n) = 0$.*

Then there exists a (unique) $L^p_{\mathbf{B}}$-stochastic measure x such that $m(g) = \int g\,dx$ for all elements g of $\mathcal{G}$; moreover, the set $x(\mathcal{Q})$ is a bounded subset of $L^p_{\mathbf{B}}$. If $p = 0$ or if **B** *is a finite-dimensional vector space, property* (i) *is necessarily fulfilled.*

(3) *Let us assume that $p \geqslant 1$ and that* **B** *is a finite-dimensional vector space; then every bounded subset of the space of all $L^p_{\mathbf{B}}$-stochastic linear functionals is relatively compact for the topology of the usual convergence when $L^p_{\mathbf{B}}$ is endowed with its weak topology $\sigma(L^p_{\mathbf{B}}, (L^p_{\mathbf{B}})')$.*

Proof (1) Statement (1) is an immediate corollary of the theorem of Section 12.7.

(2) Let m be an $L^p_{\mathbf{B}}$-stochastic linear functional. In the first step of our proof we have to establish the following property:

(j) let $(g_n)_{n>0}$ be a decreasing sequence of elements of $\mathcal{G}$ such that $g_n \downarrow 0$ and $G_n \downarrow \emptyset$, where $G_n := \{\omega : \exists t,\ g_n(\omega, t) > 0\}$; then $m(g_n)$ goes to zero for the usual topology of $L^p_{\mathbf{B}}$.

If $p = 0$, this property is obvious. Let us assume that $p > 0$ and that property (j) does not hold. Thus let us assume that $\| g_1 \| \leqslant 1$ and $\|m(g_n)\|_p \geqslant \epsilon > 0$ for all integers n, the sequence $(g_n)_{n>0}$ being as in (j) (let us recall that $\| \cdot \|_p$ is the F-norm defined in Section 12.4).

We define two increasing functions f and h from $\mathbf{N}$ into $\mathbf{N}$ (two "subsequences") as follows: $f(1) := 1 =: h(1)$, $f(n+1)$ is such that $f(n+1) \geqslant h(n+1)$, and

$$\|m(g_{h(n)})1_{F(n+1)}\|_p \leqslant \tfrac{1}{2}\epsilon,$$

where

$$F(n+1) := \{\omega \mid g_{f(n+1)} > 0\}.$$

We put

$$u_n := g_{h(n)} - g_{f(n+1)} \qquad \text{and} \qquad v_n = m(u_n).$$

We have $\|v_n\|_p \geqslant \frac{1}{2}\epsilon$. Now let $h(n+1)$ be an integer such that $h(n+1) \geqslant f(n+1)$ and

$$\|v_n 1_{H(n+1)}\|_p \leqslant \epsilon 4^{-n},$$

where

$$H(n+1) := \{\omega : g_{h(n+1)} > 0\}.$$

Now the functions f and h being defined, we put

$$w_n := \sum_{k=1}^{n} v_n.$$

On the one hand, we have

$$w_n = m\left(\sum_{k=1}^{n} u_k\right) \qquad \text{with} \quad 0 \leqslant \sum_{k=1}^{n} u_k \leqslant h_1.$$

On the other hand, if we put $J(k) := H(k) \backslash H(k+1)$, we have for every integer k such that $k < n$,

$$\|w_n 1_{J(k)}\|_p \geqslant \|v_k 1_{J(k)}\|_p - \sum_{j=1}^{k-1} \|v_j 1_{J(k)}\|_p$$

$$\geqslant \frac{1}{2}\epsilon - \sum_{j=1}^{k} \epsilon 4^{-j} \geqslant \frac{1}{6}\epsilon,$$

which implies

$$\lim_{n\to\infty} \|w_n\|_p = +\infty,$$

and this contradicts the continuity of the mapping m, which completes the proof of property (j).

(3) Now we prove (2) of the theorem. Let us assume that m is an $L^p_{\mathbf{B}}$-stochastic linear functional. Let $(f_n)_{n>0}$ be a decreasing sequence of elements of $\mathcal{G}$ such that $f_n \downarrow 0$, and let $\epsilon > 0$; for every integer n we put $G_n := \{\omega : \exists t \text{ with } f_n(\omega, t) > \epsilon\}$ and $g_n := f_n \vee \epsilon - \epsilon$. We have $G_n = \{\omega : \exists t$ with $g_n(\omega, t) > 0\}$, $G_n \downarrow \emptyset$, and $g_n \downarrow 0$; then (see (2)(j) above) $\lim_{n\to\infty} m(g_n) = 0$.

But, we have

$$m(f_n) = m(g_n) + m(f_n - f_n \vee \epsilon + \epsilon) \qquad \text{with} \quad \|f_n - f_n \vee \epsilon + \epsilon\| \leqslant \epsilon.$$

According to the continuity of m, this implies that $m(f_n)$ goes to zero when n goes to infinity.

(4) According to (3) above and property (2)(i) in the theorem, there exists a Daniell extension for m (see Section 12.6), and there exists an $L^p_{\mathbf{B}}$-stochastic measure x such that $m(g) = \int g\,dx$ for all elements g of $\mathcal{G}$.

(5) When **B** is a finite-dimensional vector space, according to [MaO], property (2)(i) of the theorem is necessarily fulfilled (see (1) in the proof of the theorem of Section 12.7).

(6) When **B** is a finite-dimensional vector space, every bounded subset of $\mathcal{L}(\mathcal{G}, L^p_{\mathbf{B}})$ is relatively compact for pointwise convergence, $L^p_{\mathbf{B}}$ being endowed with its weak topology (see [Bou] or [Tre]). Thus since the space of all the $L^p_{\mathbf{B}}$-stochastic linear functionals is clearly a closed subspace of $\mathcal{L}(\mathcal{G}, L^p_{\mathbf{B}})$ for this topology, this implies (3) of the theorem.

13.4 Compactness Property for Sets of Stochastic Measures

Theorem *Let us assume that* **B** := **R** *and that Ω is a compact space. Let $\mathcal{K}$ be the space of all real continuous functions on Ω. We denote by $\mathcal{M}$ the space of all real bilinear functions on $(\mathcal{G} \times \mathcal{K})$ that are continuous (for the uniform topology of $\mathcal{G}$ and $\mathcal{K}$) and satisfy the following property:*

(i) *if g belongs to $\mathcal{G}$, F belongs to $\mathcal{F}$, and $g 1_F = g$, the Radon measure (on Ω), $m(g, \cdot)$ is null on $(\Omega \backslash F)$.*

An element m of $\mathcal{M}$ will be said to be dominated by P if for every element g of $\mathcal{G}$ the Radon measure $m(g, \cdot)$ is dominated by P.

(1) *Let x be an* $L^1_{\mathbf{R}}$*-stochastic measure. Let m be the real function defined on* $(\mathcal{G} \times \mathcal{K})$ *by*

$$m(g,k) = E\left[k\int g\,dx\right];$$

then m belongs to $\mathcal{M}$ *and is dominated by P.*

(2) *Let m be an element of* $\mathcal{M}$ *that is dominated by P. Then there exists a (unique)* $L^1_{\mathbf{R}}$*-stochastic measure such that for all the elements* (g,k) *of* $(\mathcal{G} \times \mathcal{K})$ *we have*

$$m(g,k) = E\left[k\int g\,dx\right].$$

(3) *Every bounded subset of* $\mathcal{M}$ *is relatively compact for the topology of pointwise convergence.*

Proof The theorem of Section 13.3 implies (1). Now let m be an element of $\mathcal{M}$. Let $\mathcal{K}'$ be the strong topological dual of $\mathcal{K}$: m induces a mapping $\overline{m}$ from $\mathcal{G}$ into $\mathcal{K}'$, defined by $\langle \overline{m}(g), k\rangle = m(g,k)$, for all elements k of $\mathcal{K}$.

This mapping $\overline{m}$ is weakly continuous; thus it is strongly continuous (cf. [Bou-2] or [Tre]). Thus it can be considered as a continuous mapping from $\mathcal{G}$ into $L_1(\Omega, \mathcal{F}, P)$. Now we can use the theorem of Section 13.3, which proves (2).

The proof of (3) is quite similar to the proof of (3) in Section 13.3.

STOCHASTIC RADON MEASURE

13.5 Hypotheses and Notation

We consider the hypotheses and notation given in Section 12.1 in the following particular setting:

T is a metrizable compact space;

$\mathcal{J}$ is a boolean semialgebra of subsets of T, the interiors of the elements of $\mathcal{J}$ being a base of open sets for T;

$(\Omega, \mathcal{F}, P)$ is a complete probabilized space.

For every element t of T, $\mathcal{F}_t$ is a sub-σ-algebra of $\mathcal{F}$ that includes all P-null sets of $\mathcal{F}$; for every element J of $\mathcal{J}$, $\mathcal{F}_J$ is defined by $\mathcal{F}_J = \bigcap_{t\in J}\mathcal{F}_t$.

The following property is assumed to be satisfied:

(i) for every element J of $\mathcal{J}$ there exists an element t of $\bar{J}$ such that $\mathcal{F}_t \subset \mathcal{F}_j$.

Such a point t will be called an initial point for J (such an initial point is not necessarily unique).

Of course such hypotheses generalize the setting considered in Chapter 1 of this book; for other examples see [Met-5].

13.6 Adapted and Predictable Processes

Definition Let X be a real process, i.e., a mapping from $(\Omega \times T)$ into $\mathbf{R}$. Such a process will be called adapted if X_t is $\mathcal{F}_t$-measurable for all elements t of T.

$\mathcal{R}$, $\mathcal{A}$, and P' are defined as above (see Sections 1 and 12.1); the following proposition is easy to verify.

Proposition *Every predictable process is an adapted process. Every continuous process (i.e., such that the mapping $t \rightsquigarrow X(\omega, t)$ is continuous for every element ω of Ω) that is adapted is a predictable process. The σ-algebra of predictable sets is generated by all such continuous adapted processes.*

13.7 Radon Stochastic Measure

We denote by $\mathcal{G}$ the vector space of all uniformly bounded adapted continuous processes; this space $\mathcal{G}$ will be endowed with its uniform norm

$$\|X\| = \sup_{\omega, t} |X(\omega, t)|.$$

An $L^p_{\mathbf{B}}$-valued function m on $\mathcal{G}$ will be called a Radon stochastic measure if it is a linear continuous mapping (for the uniform topology of $\mathcal{G}$ and the usual topology of $L^p_{\mathbf{B}}$) that satisfies the following property:

(i) if g belongs to $\mathcal{G}$, F belongs to $\mathcal{F}$, and $1_F g$ belongs to $\mathcal{G}$, we have

$$m(1_F g) = 1_F m(1_F g).$$

13.8 Isometric Representation

Theorem *Let x be an $L^p_{\mathbf{B}}$-stochastic measure; we denote by m_x the linear mapping from $\mathcal{G}$ into $L^p_{\mathbf{B}}$, defined by*

$$m_x(g) := \int g\,dx \qquad \textit{for every element } g \textit{ of } \mathcal{G}.$$

Then m_x is an $L^p_{\mathbf{B}}$-Radon stochastic measure. The mapping $x \rightsquigarrow m_x$ is an algebraic isometry from the vector space of all $L^p_{\mathbf{B}}$-stochastic measures onto the vector space of all $L^p_{\mathbf{B}}$-Radon stochastic measures.

Proof This theorem is clearly a corollary of the theorem of Section 13.3.

HISTORICAL NOTES

Section 12. Viewing stochastic integration as a particular case of integration, with respect to vector measures was proposed by the authors in a series of papers (for example [Met-1], [Met-2], [Pel-3], [Pel-4]). The book by Kussmaul is devoted to the exposition of this theory ([Kus]) in the case of real processes. The general framework considered here, in Sections 12.1–12.11, where the index set T is allowed to be a quite general topological space and where stochastic measures take their values in $L^0(\Omega, \mathcal{F}, P)$, is the one adopted by the authors in [MeP-4].

The deep Dellacherie–Meyer–Mokobodski theorem, which is actually a direct consequence of a theorem on bounded sets in L_0 (the theorem of Section 12.11) has not been published yet in a formal way by its authors. The same theorem for stochastic measures with values in $L^1(\Omega, \mathcal{F}, P)$ was treated first in [Met-5].

Section 13. This section reproduces essentially the corresponding part of the authors' paper [MeP-4].

CHAPTER 6

SPECIAL FEATURES OF INFINITE-DIMENSIONAL STOCHASTIC INTEGRATION

Up until this chapter we have been concerned with problems that can be formulated in exactly the same way for finite-dimensional and infinite-dimensional processes. In several instances the infinite-dimensional case presented a few peculiarities (for example, the relation between π^*-processes and semimartingales), but the techniques involved were essentially the same.

This chapter is devoted to problems specific to the infinite-dimensional case, which arise mainly from two necessities: on the one hand it is impossible to escape considering unbounded operators and therefore unbounded operator valued processes (see [Cur], [MPi-2]); on the other hand perturbations such as "white noise in space" have been introduced in the study of stochastic distributed systems ([Ben], [Par-2], [Vio-2]). But this "white noise" is the stochastic integral operator associated with the standard cylindrical brownian motion as defined in Section 15.5, which is not an ordinary sense process; it will appear in this chapter as a particular "cylindrical martingale."

The leading idea of this chapter is to look for spaces of processes and topologies on those spaces for which the stochastic integral is an isometry. We see in this way how unbounded operator valued processes naturally come in. In Section 14 the isometric stochastic integral is built with respect to a Hilbert-valued square integrable martingale. In Section 15 cylindrical processes are introduced while in Section 16 the isometric stochastic integral with respect to 2-cylindrical martingales is constructed.

While the first four chapters involve only elementary material from measure and integration theory and can be considered to be self-contained, this last chapter requires some amount of functional analysis. We refer when necessary to some basic books listed in the bibliography when results on operators, weakly measurable functions, lifting theorems, etc. are needed.

14 THE ISOMETRIC INTEGRAL OF A HILBERT-VALUED SQUARE INTEGRABLE MARTINGALE

As in previous sections, a stochastic basis $(\Omega, (\mathcal{F}_t), P)_{t \in T}$ is given, and we make the usual assumptions of completeness and right-continuity on the $\mathcal{F}_t$s.

14.1 Introduction

It has been noticed in Section 2.6 that when M is a real square integrable martingale, the mapping $Y \rightsquigarrow \int_{[0, t_m]} Y\,dM$ is an isometry from $L^2(\Omega', \mathcal{P}, \alpha)$ into $L^2(\Omega, \mathcal{F}_{t_m}, P)$, where α denotes the Doléans measure of $|M|^2$. As an immediate consequence, $Y \rightsquigarrow (\int Y\,dM)$ is an isometry from $L^2(\Omega', \mathcal{P}, \alpha)$ into $\mathfrak{M}^2_{[0, t_m]}$, the Hilbert space of real square integrable martingales on $[0, t_m]$.

In the case of a Hilbert-valued martingale M, instead of an isometry, we have a contraction. This so to speak expresses that the norm in $L^2(\Omega', \mathcal{P}, \alpha)$ is "too big," and we may lose many potentially integrable processes by taking too small a closure of $\mathcal{E}(\mathbf{H})$.

This is the problem we investigate in this section. We shall define a norm on the space of processes, which will lead to an isometry property for the mapping $Y \rightsquigarrow (\int Y\,dM)$. The set of integrable processes will increase considerably: in the infinite-dimensional case it will contain, in particular, processes whose values are unbounded operators. This section contains materials originating from [MPi-2].

14.2 More on Linear Operators and Tensor Products

We review a few concepts which will be used freely throughout the chapter. For proofs and details the reader is referred, for example, to [Tre] or [GeW].

(1) $\mathbf{H} \hat{\otimes}_2 \mathbf{H}$ *and Hilbert–Schmidt Operators.* Let $\mathbf{H}$ be a separable real Hilbert space with scalar product noted xy as in Section 10 or $\langle x, y \rangle_{\mathbf{H}}$ when $\mathbf{H}$ has to be specified. We have defined in Section 3.4 the Hilbert–Schmidt tensor product space $\mathbf{H} \hat{\otimes}_2 \mathbf{H}$. We write $(\ |\ )_2$ for the scalar product and $\|\cdot\|_2$ for the norm in this Hilbert space.

For every $b \in \mathbf{H} \hat{\otimes}_2 \mathbf{H}$ it is possible to associate the bilinear form $(h, g) \rightsquigarrow (b \mid h \otimes g)_2$. The tensor b is said to be *positive* if this bilinear form is such that $(b \mid h \otimes h)_2 \geqslant 0$ for all $h \in \mathbf{H}$, and it is said to be *symmetric* if $(b \mid h \otimes g)_2 = (b \mid g \otimes h)_2$ for all g and h in $\mathbf{H}$.

In this way $\mathbf{H} \hat{\otimes}_2 \mathbf{H}$ can be considered as a particular set of continuous bilinear forms on $\mathbf{H} \times \mathbf{H}$. To every bilinear form b on $\mathbf{H} \times \mathbf{H}$ is associated in a one-to-one way a continuous linear operator $\tilde{b}$ through the formula

$$b(h, g) = \tilde{b}(h)g \qquad \text{for all} \quad h, g \in \mathbf{H}.$$

The linear operator $\tilde{b}$ is called a *Hilbert–Schmidt operator* if $b \in \mathbf{H} \hat{\otimes}_2 \mathbf{H}$ and $\|b\|_2$ is its Hilbert–Schmidt norm. It is called self-adjoint if b is symmetric.

If we consider an orthonormal basis (h_n) of $\mathbf{H}$, we have the following useful characterization of Hilbert–Schmidt operators: *$\tilde{b}$ is a Hilbert–Schmidt operator iff $\sum_n \|\tilde{b}(h_n)\|^2 < \infty$, and moreover, $\|\tilde{b}\|_2^2 = \sum_n \|\tilde{b}(h_n)\|^2$* (see, for example, [GeW]).

For every self-adjoint Hilbert–Schmidt operator $\tilde{b}$ in $\mathbf{H}$ there exists an orthogonal basis (h_n) in $\mathbf{H}$ such that

$$\tilde{b}(h) = \sum_n \lambda_n (hh_n) h_n \qquad \forall h \in \mathbf{H}, \tag{14.2.1}$$

where the λ_n are real numbers called the eigenvalues of $\tilde{b}$ and are such that

$$\sum_n \lambda_n^2 = \|\tilde{b}\|_2^2 < \infty.$$

These eigenvalues are positive if $\tilde{b}$ is positive.

The conjugate $\tilde{b}^*$ of a Hilbert–Schmidt operator, i.e., the operator $\tilde{b}^*$ defined by $\tilde{b}(h)g = h\tilde{b}^*(g)$, is equally Hilbert–Schmidt with $\|\tilde{b}^*\|_2 = \|\tilde{b}\|_2$.

We denote by $\mathcal{L}_2(\mathbf{H}; \mathbf{H})$ the Hilbert space of Hilbert–Schmidt operators. From the preceding the scalar product in $\mathcal{L}_2(\mathbf{H}; \mathbf{H})$ is expressed by

$$(\tilde{b}_1 \mid \tilde{b}_2) = \|\tilde{b}_1 \circ \tilde{b}_2^*\|.$$

Starting from the formula $(h \otimes g \mid h' \otimes g') = (hh')(gg')$, it is possible, in an entirely analogous way, to define the space $\mathbf{H} \hat{\otimes}_2 \mathbf{G}$ and from there on the space $\mathcal{L}_2(\mathbf{H}; \mathbf{G})$ of Hilbert–Schmidt operators from $\mathbf{H}$ into $\mathbf{G}$. Those operators $\tilde{b}$ are still characterized by the property

$$\sum_n \|\tilde{b}(h_n)\|^2 = \|\tilde{b}\|_2^2 < \infty$$

for every orthogonal basis (h_n) in $\mathbf{H}$.

(2) $\mathbf{B} \hat{\otimes}_1 \mathbf{B}$ *for a Banach space* $\mathbf{B}$. In section 3.9 we recalled the existence of the tensor product space $\mathbf{B} \hat{\otimes}_1 \mathbf{B}$, which can be defined for every Banach space $\mathbf{B}$ as the completion of $\mathbf{B} \otimes \mathbf{B}$ for a norm (written $\|\cdot\|_1$) that has the following property: for every Banach space $\mathbf{G}$ and every continuous bilinear mapping b from $\mathbf{B} \times \mathbf{B}$ into $\mathbf{G}$, there exists a unique b continuous (for the norm $\|\cdot\|_1$) linear mapping $\bar{b}: \mathbf{B} \otimes \mathbf{B} \to \mathbf{G}$ such that

$b(x, y) = \bar{b}(x \otimes y)$.

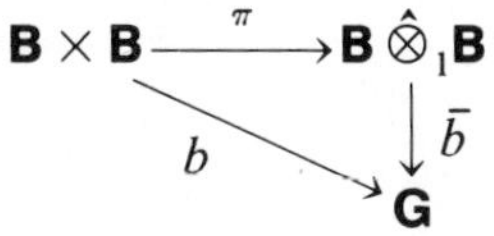

In other words every bilinear mapping b can be factored as $\bar{b} \circ \pi$, where $\pi(x, y) = x \otimes y$ and $\bar{b}$ is continuous on $\mathbf{B} \hat{\otimes}_1 \mathbf{B}$.

Taking $\mathbf{G} = \mathbf{R}$, we see in particular that the dual $(\mathbf{B} \hat{\otimes}_1 \mathbf{B})'$ of $\mathbf{B} \hat{\otimes}_1 \mathbf{B}$ is isomorphic to the Banach space of continuous bilinear forms, with the usual norm.

For a Hilbert space $\mathbf{H}$ there is a continuous linear injection from $\mathbf{H} \hat{\otimes}_1 \mathbf{H}$ into $\mathbf{H} \hat{\otimes}_2 \mathbf{H}$ that extends the identical mapping $h \otimes g \rightarrow h \otimes g$. Actually $\|b\|_2 \leqslant \|b\|_1$ for every $b \in \mathbf{H} \hat{\otimes}_1 \mathbf{H}$, with equality for every b, which can be written $b = h \otimes g$. In this last situation we have $\|h \otimes g\|_1 = \|h \otimes g\|_2 = \|h\|_{\mathbf{H}} \| g\|_{\mathbf{H}}$.

We recall also that the linear form trace on $\mathbf{H} \hat{\otimes}_1 \mathbf{H}$ is defined as the unique continuous linear extension to $\mathbf{H} \hat{\otimes}_1 \mathbf{H}$ of the mapping $h \otimes g \rightarrow hg$.

Let us write $\langle x, x' \rangle$ for the canonical bilinear form on $\mathbf{B} \times \mathbf{B}'$. From the definition of $\mathbf{B} \hat{\otimes}_1 \mathbf{B}$ and $\mathbf{B}' \hat{\otimes}_1 \mathbf{B}'$ it is immediately seen that the mapping $(x \otimes y, x' \otimes y') \rightarrow \langle x, x' \rangle \langle y, y' \rangle$ extends uniquely into a continuous bilinear form $\langle \cdot, \cdot \rangle$ on $(\mathbf{B} \hat{\otimes}_1 \mathbf{B}) \times (\mathbf{B}' \hat{\otimes}_1 \mathbf{B}')$.

Thus $\mathbf{B} \hat{\otimes}_1 \mathbf{B}$ appears to be isomorphic to a subspace of the dual space $(\mathbf{B}' \hat{\otimes}_1 \mathbf{B}')'$.

On the one hand this gives a meaning to the notion of positive and symmetric elements of $\mathbf{B} \hat{\otimes}_1 \mathbf{B}$: the element $b \in \mathbf{B} \hat{\otimes}_1 \mathbf{B}$ is said to be *positive* if $\langle b, x \otimes x \rangle \geqslant 0$ for all $x \in \mathbf{B}$ and *symmetric* if $\langle b, x \otimes y \rangle = \langle b, y \otimes x \rangle$ for all $x, y \in \mathbf{B}'$.

On the other hand we can associate with every element b of $\mathbf{B} \hat{\otimes}_1 \mathbf{B}$ a continuous linear operator $\tilde{b}$ from $\mathbf{B}'$ into $\mathbf{B}$, uniquely defined by

$$\langle \tilde{b}(h), g \rangle = \langle b, h \otimes g \rangle$$

when $\mathbf{B}$ is reflexive. Such an operator is called *nuclear*.

At last, if q_1 and q_2 are elements of $\mathfrak{L}(\mathbf{B}; \mathbf{K})$, $q_1 \otimes q_2$ denotes the element of $\mathfrak{L}(\mathbf{B} \hat{\otimes}_1 \mathbf{B}; \mathbf{K} \hat{\otimes}_1 \mathbf{K})$ that is the unique continuous linear extension of $x \otimes y \rightarrow q_1(x) \otimes q_2(y)$.

(3) *Nuclear Operators in Hilbert Spaces.* Identifying $\mathbf{H}'$ with $\mathbf{H}$ as usual when $\mathbf{H}$ is a Hilbert space, we see that $\tilde{b} \in \mathfrak{L}(\mathbf{H}'; \mathbf{H})$ is a *nuclear operator* iff the linear form $x \otimes g \rightarrow \tilde{b}hg$ can be identified (in the sense of (1) above) with an element b of $\mathbf{H} \hat{\otimes}_1 \mathbf{H}$.

Self-adjoint nuclear operators are characterized by the following property: $\tilde{b}$ is a self-adjoint nuclear operator iff there exists an orthonormal basis (h_n) such that

$$\tilde{b}(h) = \sum_n \lambda_n (hh_n)h_n, \qquad \forall h \in \mathbf{H}, \tag{14.2.2}$$

where the λ_ns are real numbers, called eigenvalues, such that $\sum_n |\lambda_n| \lesssim \infty$.

One can check that if b is the element of $\mathbf{H} \hat{\otimes}_1 \mathbf{H}$ associated with $\tilde{b}$, we have

$$\mathrm{Tr}(\tilde{b}) = \sum_n \lambda_n \tag{14.2.3}$$

and

$$\|\tilde{b}\|_1 = \sum_n |\lambda_n|. \tag{14.2.4}$$

Formulas (14.2.2) and (14.2.4) show that every positive self-adjoint operator $\tilde{b}$ can be written as

$$\tilde{b} = \tilde{q} \circ \tilde{q}, \qquad \tilde{q} \in \mathcal{L}_2(\mathbf{H}; \mathbf{H}),$$

the eigenvalues of $\tilde{q}$ being the square roots of those of $\tilde{b}$. In this case we write

$$\tilde{q} = \tilde{b}^{1/2}. \tag{14.2.5}$$

We shall write $\mathcal{L}_1(\mathbf{H}; \mathbf{H})$ for the space of nuclear operators in $\mathbf{H}$.

We mention finally the following properties:

(a) if $q \in \mathcal{L}_2(\mathbf{H}; \mathbf{H})$, then $q \circ q \in \mathcal{L}_1(\mathbf{H}; \mathbf{H})$ and $\|q\|_1 = \|q\|_2^2$;

(b) if $q \in \mathcal{L}_2(\mathbf{H}; \mathbf{H})$ and $u \in \mathcal{L}(\mathbf{H}; \mathbf{H})$, then $q \circ u \in \mathcal{L}_2(\mathbf{H}; \mathbf{H})$ and $\|q \circ u\|_2 \leqslant \|q\|_2 \|u\|$;

(c) if $q \in \mathcal{L}_2(\mathbf{H}; \mathbf{H})$ (resp. $q \in \mathcal{L}_1(\mathbf{H}; \mathbf{H})$), then the adjoint q^* is an element of $\mathcal{L}_2(\mathbf{H}; \mathbf{H})$ (resp. $\mathcal{L}_1(\mathbf{H}; \mathbf{H})$) with the same norm.

14.3 The Processes Q_M and $\langle M \rangle$ of a Square Integrable Martingale

With an $\mathbf{H}$-valued square integrable martingale, we have associated the Doléans measure α_M of $\|M\|^2$, which is a dominating measure for M in the sense of Section 2.3.

With M we associate also the Doléans function of $M \otimes M$, that is,

$$d_{M \otimes M}(F \times]s, t]) = E\left\{1_F\left(M_t^{\otimes_2} - M_s^{\otimes_2}\right)\right\} \in \mathbf{H} \hat{\otimes}_2 \mathbf{H} \tag{14.3.1}$$

for every predictable rectangle $]s, t] \times F$, or as shown in Section 4.4,

$$d_{M \otimes M}(F \times]s, t]) = E\left\{1_F(M_t - M_s)^{\otimes_2}\right\} \in \mathbf{H} \hat{\otimes}_2 \mathbf{H}. \tag{14.3.2}$$

Since the linear form "Trace" on $\mathbf{H}\hat{\otimes}_2\mathbf{H}$ is the linear continuous extension of the mapping $(x, y) \rightsquigarrow \operatorname{Tr}(x \otimes y) = xy$, it is clear from the above formula that

$$\alpha_M = \operatorname{Tr} d_{M \otimes M}.$$

The existence of a σ-additive extension of $d_{M\otimes M}$ is therefore trivial. If μ_M denotes this extension, we have

$$\alpha_M = \operatorname{Tr} \mu_M.$$

From the inequality

$$\| \mu_M(F \times]s,t])\|_{\mathbf{H}\hat{\otimes}_2\mathbf{H}} \leqslant E\left(\|1_F(M_t - M_s)^{\otimes_2}\|_{\mathbf{H}\hat{\otimes}_2\mathbf{H}}\right)$$

we have

$$\| \mu_M(F \times]s,t])\|_{\mathbf{H}\hat{\otimes}_2\mathbf{H}} \leqslant \alpha_M(F \times]s,t]), \tag{14.3.3}$$

and we easily deduce that the variation of the vector-valued measure μ_M is smaller than α_M.

As already done in Section 9.14, we may use a Radon–Nikodym theorem for Hilbert-valued measures to obtain immediately the existence of a $\mathbf{H}\hat{\otimes}_2\mathbf{H}$-valued predictable process Q_M such that for every predictable G

$$\mu_M(G) = \int_G Q_M \, d\alpha_M,$$

with $\|Q_M\| \leqslant 1$ in view of (14.3.3).

More in fact can be said: since

$$\|(M_t - M_s)^{\otimes_2}\|_{\mathbf{H}\hat{\otimes}_2\mathbf{H}} = \|(M_t - M_s)^{\otimes_2}\|_{\mathbf{H}\hat{\otimes}_1\mathbf{H}} = \|M_t - M_s\|^2_{\mathbf{H}},$$

inequality (14.3.3) holds for the nuclear norm, which is a stronger inequality. In this case it is more precisely an equality because from the definition, $\mu_M(]s,t] \times F)$ is clearly a positive element in $\mathbf{H}\hat{\otimes}_1\mathbf{H}$ and

$$\| \mu_M(F \times]s,t])\|_{\mathbf{H}\hat{\otimes}_1\mathbf{H}} = \operatorname{Tr}\| \mu_M(F \times]s,t])\|.$$

If we are ready to admit that the Radon–Nikodym theorem holds for $\mathbf{H}\hat{\otimes}_1\mathbf{H}$-valued measures with bounded variation as a consequence of the Shatten theorem ([Tre]), which states that $\mathbf{H}\hat{\otimes}_1\mathbf{H}$ is a separable dual space of a Banach space, we obtain the first part of the following theorem.

Theorem (1) *There is one predictable $\mathbf{H}\hat{\otimes}_1\mathbf{H}$-valued process Q_M, defined up to α_M-equivalence such that for every $G \in \mathcal{P}$*

$$\mu_M(G) = \int_G Q_M \, d\alpha_M.$$

Moreover, Q_M takes its values in the set of positive symmetric elements of $\mathbf{H}\hat{\otimes}_1\mathbf{H}$ *and*

$$\operatorname{Tr} Q_M(\omega, s) = \|Q_M(\omega, s)\|_{\mathbf{H}\hat{\otimes}_1\mathbf{H}} = 1, \qquad \alpha_M \text{ a.e.} \tag{14.3.4}$$

(2) *The process*

$$\langle M \rangle_t := \int_{[0,t]} Q_M \, d\langle M \rangle$$

has finite variation, is predictable, admits μ_M as its Doléans measure, and is such that $M^{\otimes_2} - \langle M \rangle$ is a martingale.

Proof (1) The existence of Q_M has been proved above. Since μ_M takes its values in the set of positive symmetric elements of $\mathbf{H}\hat{\otimes}_1\mathbf{H}$, the same holds for Q_M. The equality (14.3.4) then follows immediately from this and from the fact that $\alpha_M = \operatorname{Tr} \mu_M$.

(2) The second part of the theorem follows immediately from Section 9.14.

14.4 The Space $L^*(\mathbf{H}; \mathbf{G}; \mathcal{P}, M)$

With the $\mathbf{H}\hat{\otimes}_1\mathbf{H}$-valued process Q_M we associate the $\mathcal{L}^1(\mathbf{H}; \mathbf{H})$-valued process $\tilde{Q}_M$ related to Q_M by

$$\tilde{Q}_M h g = Q_M(h \otimes g), \qquad (h, g) \in \mathbf{H} \times \mathbf{H}.$$

We may speak of the square root $\tilde{Q}_M^{1/2}$ of $\tilde{Q}_M$, which is a Hilbert–Schmidt operator valued process.

Let $\mathbf{G}$ be another separable Hilbert space.

We call $L^*(\mathbf{H}; \mathbf{G}; \mathcal{P}, M)$ the space of processes X, the values of which are (possibly noncontinuous) linear operators from $\mathbf{H}$ into $\mathbf{G}$, with the following properties:

(i) the domain $\mathcal{D}X(\omega, t)$ of $X(\omega, t)$ contains $\tilde{Q}_M^{1/2}(\omega, t)(\mathbf{H})$ for every (ω, t);

(ii) for every $h \in \mathbf{H}$, the $\mathbf{G}$-valued process $X \circ \tilde{Q}_M^{1/2}(h)$ is predictable;

(iii) for every $(\omega, t) \in \Omega'$, $X(\omega, t) \circ \tilde{Q}_M^{1/2}(\omega, t)$ is a Hilbert–Schmidt operator and

$$\int_{\Omega'} \|X \circ \tilde{Q}_M^{1/2}\|_2^2 \, d\alpha_M < \infty.$$

We then have the following proposition.

Proposition *For every $X, Y \in L^*(\mathbf{K}; \mathbf{G}; \mathcal{P}, M)$ the process $X \circ \tilde{Q}_M \circ Y^*$ is an $\mathcal{L}_1(\mathbf{H}; \mathbf{G})$-valued predictable process with $\int_{\Omega'} \operatorname{Tr}(X \circ \tilde{Q}_M \circ Y^*) \, d\alpha_M < \infty$. The bilinear form $(X, Y) \rightsquigarrow \int_{\Omega'} \operatorname{Tr}(X \circ \tilde{Q}_M \circ Y^*) \, d\alpha_M$ is a scalar product on $L^*(\mathbf{H}; \mathbf{G}; \mathcal{P}, M)$, and for this scalar product this space is complete.*

Proof We first prove that $\mathrm{Tr}(X\circ\tilde{Q}_M\circ Y^*)$ is a predictable process. Since

$$\begin{aligned}\mathrm{Tr}(X\circ\tilde{Q}_M\circ Y^*) &= \mathrm{Tr}(Y\circ\tilde{Q}_M\circ X^*)\\ &= \tfrac{1}{4}\{\mathrm{Tr}[(X+Y)\circ\tilde{Q}_M\circ(X^*+Y^*)]\\ &\quad -\mathrm{Tr}[(Y-X)\circ\tilde{Q}_M\circ(Y^*-X^*)]\},\end{aligned}$$

we have only to prove that for every $X\in L^*(\mathbf{H};\mathbf{G};\mathcal{P},M)$ the process $\mathrm{Tr}(X\circ\tilde{Q}_M\circ X^*)$ is predictable.

Let (h_n) be an orthonormal basis of $\mathbf{H}$. Since

$$\mathrm{Tr}(X\circ\tilde{Q}_M\circ X^*) = \|X\circ\tilde{Q}_M^{1/2}\|_2^2 = \sum_n \|X\circ\tilde{Q}_M^{1/2}(h_n)\|_{\mathbf{G}}^2,$$

hypothesis (ii) shows that this process is predictable. Since

$$\mathrm{Tr}(X\circ\tilde{Q}_M\circ Y^*) \leqslant \|X\circ\tilde{Q}_M^{1/2}\|_2\|Y\circ\tilde{Q}_M^{1/2}\|_2$$

and in view of the Schwarz inequality, is is immediate that

$$(X,Y)\rightsquigarrow\int_{\Omega'}\mathrm{Tr}(X\circ\tilde{Q}_M\circ Y^*)d\alpha_M$$

is a positive continuous bilinear form on $L^*(\mathbf{H};\mathbf{G};\mathcal{P},M)$.

We now show that every Cauchy sequence for this scalar product has a limit in $L^*(\mathbf{H};\mathbf{G};\mathcal{P},M)$. Let us then consider (X_n) with

$$\lim_{n,m\to\infty}\int\|(X_n-X_m)\circ\tilde{Q}_M^{1/2}\|_2^2\,d\alpha_M = 0. \tag{14.4.1}$$

In the space $L^2_{\mathcal{L}_2(\mathbf{H};\mathbf{G})}(\Omega',\mathcal{P},\alpha_M)$ the sequence $X_n\circ\tilde{Q}_M$ converges to some Y, and we can extract a subsequence $(X_{n_k})_{k\geqslant 0}$ such that

$$\lim_k X_{n_k}(\omega,t)\circ\tilde{Q}_M^{1/2}(\omega,t) = Y(\omega,t), \qquad \alpha_M \text{ a.e.}$$

Since $\tilde{Q}_M^{1/2}(\omega,t)f=0$ implies $Y(\omega,t)f=0$, it is possible to write Y as

$$Y(\omega,t) = X(\omega,t)\circ\tilde{Q}_M^{1/2}$$

for some $X(\omega,t)$ which linearly maps $\tilde{Q}_M^{1/2}(\mathbf{H})$ into $\mathbf{G}$, and which clearly meets conditions (i)–(iii). This completes the proof.

14.5 The Space $\Lambda^2(\mathbf{H};\mathbf{G};\mathcal{P},M)$

We call $\mathcal{S}(\mathcal{L}(\mathbf{H};\mathbf{G}))$ the space of $\mathcal{L}(\mathbf{H};\mathbf{G})$-valued $\mathcal{Q}$-simple processes and $\Lambda^2(\mathbf{H};\mathbf{G};\mathcal{P},M)$ the closure of $\mathcal{S}(\mathcal{L}(\mathbf{H};\mathbf{G}))$ in $L^*(\mathbf{H};\mathbf{G};\mathcal{P},M)$. We thus obtain a Hilbert subspace of $L^*(\mathbf{H};\mathbf{G};\mathcal{P},M)$.

Proposition *Every process X with the properties*

(i) *for every $(\omega,t)\in\Omega'$, $X(\omega,t)\in\mathcal{L}(\mathbf{H};\mathbf{G})$,*

(ii) *for every* $h \in \mathbf{H}$, $X(h)$ *is a predictable* $\mathbf{G}$*-valued process,*
(iii) $\int \mathrm{Tr}(X \circ \tilde{Q}_M \circ X^*)\, d\alpha_M < \infty$,

belongs to $\Lambda^2(\mathbf{H}; \mathbf{G}; \mathcal{P}, M)$.

Proof Let us assume first that X is a measurable mapping from Ω' into the Banach space $\mathcal{L}(\mathbf{H}; \mathbf{G})$ with $\sup_{\omega, t} \|X(t, \omega)\| \leqslant K < \infty$. There exists a uniformly bounded sequence (X_n) in $\mathcal{E}(\mathcal{L}(\mathbf{H}; \mathbf{H}))$ that α_M-a.e. converges to $X(\omega, t)$ in $\mathcal{L}(\mathbf{H}; \mathbf{G})$. For such a sequence one has

$$\lim_{n\to\infty} \int_{\Omega'} \|(X - X_n) \circ \tilde{Q}_M^{1/2}\|_2^2\, d\alpha_M = 0.$$

If X is a process for which (i)–(iii) hold, for a dense subset $\{h_n\}$ in the unit ball of $\mathbf{H}$ we have

$$\|X(\omega, t)\| = \sup_n \|X(\omega, t)(h_n)\|_{\mathbf{G}},$$

and $\|X\|$ is therefore predictable. For every $(\omega, s) \in \Omega'$

$$\lim_{n\to\infty} \|1_{\{\|X\| \leqslant n\}} X(\omega, s) - X(\omega, s)\| = 0,$$

and therefore

$$\lim_{n\to\infty} \|(1_{\{\|X\| \leqslant n\}} X(\omega, s) - X(\omega, s)) \circ \tilde{Q}_M^{1/2}(\omega, s)\|_2 = 0.$$

Since

$$\|(1_{\{\|X\| \leqslant n\}} X(\omega, s) - X(\omega, s)) \circ \tilde{Q}_M^{1/2}(\omega, s)\|_2^2 = 1_{\{\|X\| > n\}} \|X(\omega, s) \circ \tilde{Q}_M^{1/2}(\omega, s)\|_2^2$$
$$\leqslant \|X(\omega, s) \circ \tilde{Q}_M(\omega, s)\|_2^2,$$

the following equality holds:

$$\lim_{n\to\infty} \int \|(1_{\{\|X\| \leqslant n\}} X - X) \circ \tilde{Q}_M^{1/2}\|_2^2\, d\alpha_M = 0.$$

We have thus reduced the proof of the proposition to show that every process X with properties (i)–(iii), and which is bounded in the norm by some constant K, is in $\Lambda^2(\mathbf{H}; \mathbf{G}; \mathcal{P}, M)$.

Let (h_i) (resp. (g_i)) be an orthonormal basis in $\mathbf{H}$ (resp. in $\mathbf{G}$). We call Π_1^n (resp. Π_2^n) the orthogonal projections from $\mathbf{H}$ (resp. from $\mathbf{G}$) onto the subspaces generated by $\{h_1 \cdots h_n\}$ (resp. $\{g_1 \cdots g_n\}$). We set

$$X_n := \Pi_2^n \circ X \circ \Pi_1^n.$$

For every i we may write simultaneously

$$\lim_n \|(\Pi_2^n \circ X \circ \Pi_1^n - X) \circ \tilde{Q}_M^{1/2}(h_i)\|_{\mathbf{G}}^2 = 0$$

$$\|(\Pi_2^n \circ X \circ \Pi_1^n - X) \circ \tilde{Q}_M^{1/2}(h_i)\|_{\mathbf{G}}^2 \leqslant 4K^2 \|\tilde{Q}_M^{1/2}(h_i)\|_{\mathbf{H}}^2$$

and

$$\sum_i \|\tilde{Q}_M^{1/2}(h_i)\|_{\mathbf{H}}^2 = \|\tilde{Q}_M^{1/2}\|_2^2 < \infty.$$

From these three relations we derive

$$\lim_n \sum_i \|(\Pi_2^n \circ X \circ \Pi_1^n - X) \circ \tilde{Q}_M^{1/2}(h_i)\|_{\mathbf{G}}^2 = \lim_n \|(X_n - X) \circ \tilde{Q}_M^{1/2}\|_2^2 = 0$$

with

$$\|(X_n - X) \circ \tilde{Q}_M^{1/2}\|_2^2 \leqslant 4K^2 \|X \circ \tilde{Q}_M^{1/2}\|_2^2.$$

Therefore

$$\lim_{n \to \infty} \int \|(X_n - X) \circ \tilde{Q}_M^{1/2}\|_2^2 \, d\alpha_M = 0.$$

Since the processes X_n are in $\Lambda^2(\mathbf{H}; \mathbf{G}; \mathcal{P}, M)$, according to the beginning of the proof, we have thus proved that X also belongs to this space.

Remark and Example We should remark that $\Lambda^2(\mathbf{H}; \mathbf{G}; \mathcal{P}, M)$ contains in general other processes than those fulfilling (i)–(iii), in particular, processes whose values are unbounded operators. This appears in the following very simple example: Let (h_i) and (g_i) be orthonormal bases in $\mathbf{H}$ and $\mathbf{G}$ as above. It is easy to define a process M with independent increments such that Q is nonrandom:

$$Q(\omega, s) = \sum_{i=1}^{\infty} (1/2^i) h_i \otimes h_i \qquad \text{for every} \quad (\omega, s)$$

(for example: the brownian motion associated with Q: see Section 4.11). Let us then consider the deterministic processes X_n:

$$X_n(h) := \sum_{i=1}^{n} \sqrt{i}\,(h \otimes h_i) g_i.$$

Clearly

$$\|(X_n - X_{n+k}) \circ \tilde{Q}_M^{1/2}\|_2^2 = \sum_{i=n+1}^{n+k} \frac{i}{2^i}.$$

The sequence (X_n) is therefore a Cauchy sequence in $\Lambda^2(\mathbf{H}; \mathbf{G}; \mathcal{P}, M)$. Considering then a subsequence (X_{n_k}) such that $X_{n_k} \circ \tilde{Q}_M^{1/2}$ converges to $X \circ \tilde{Q}_M^{1/2}$ in $\mathcal{L}_2(\mathbf{H}; \mathbf{G})$, we obtain

$$\lim_k \frac{1}{2^i} X_{n_k}(h_i) = \frac{1}{2^i} X(h_i),$$

and as a consequence,

$$X(h_i) = \sqrt{i}\; g_i.$$

The operator X is surely not bounded!

14.6 The Isometric Stochastic Integral

Theorem *Let M be an element of $\mathfrak{M}^2_{[0,t_m]}(\mathbf{H})$. There exists a unique isometric linear mapping from $\Lambda^2(\mathbf{H};\mathbf{G};\mathcal{P},M)$ into $\mathfrak{M}^2_{[0,t_m]}(\mathbf{G})$ such that the image of $X := 1_{F\times]r,s]}u$ for every predictable rectangle $F\times]r,s]$ and $u \in \mathfrak{L}(\mathbf{H};\mathbf{G})$ is the martingale $(1_F[u(M_{s\wedge t}) - u(M_{r\wedge t})])_{t\in[0,t_m]}$.*

Proof We clearly have only to prove that the mapping

$$X := \sum_{i=1}^{n} 1_{F_i\times]r_i,s_i]}u_i \curvearrowright \left(\sum_{i=1}^{n} 1_{F_i}\left[u(M_{s_i\wedge t}) - u(M_{r_i\wedge t})\right]\right)_{t\in[0,t_m]}$$

is an isometric mapping from $\mathcal{E}(\mathfrak{L}(\mathbf{H};\mathbf{G}))$ into $\mathfrak{M}^2_{[0,t_m]}(\mathbf{G})$. It is always possible to assume that the predictable rectangles $F_i\times]r_i,s_i]$ are disjoint. We can then write

$$\begin{aligned}
E\left(\left\|\sum_{k=1}^{n} 1_{F_i}\left[u_i(M_{s_i}) - u_i(M_{r_i})\right]\right\|_{\mathbf{G}}^2\right) &= E\left(\sum_{i=1}^{n} \|1_{F_i}u_i(M_{s_i} - M_{r_i})\|_{\mathbf{G}}^2\right)\\
&= E\left(\sum_{i=1}^{n} 1_{F_i}\operatorname{Tr}\left[u_i\otimes u_i(M_{s_i} - M_{r_i})^{\otimes_2}\right]\right)\\
&= \operatorname{Tr}\left(\sum_{i=1}^{n} u_i\otimes u_i E\left[1_{F_i}(M_{s_i} - M_{r_i})^{\otimes_2}\right]\right)\\
&= \sum_{i=1}^{n} \operatorname{Tr}\left[u_i\otimes u_i\left(\int_{F_i\times]r_i,s_i]} \tilde{Q}_M\, d\alpha_M\right)\right]\\
&= \sum_{i=1}^{n} \int_{F_i\times]r_i,s_i]} \operatorname{Tr}(u_i\circ\tilde{Q}_M\circ u_i^*)\,d\alpha_M\\
&= \int \operatorname{Tr}(X\circ\tilde{Q}_M\circ X^*)\,d\alpha_M.
\end{aligned}$$

This proves the theorem.

The image in $\mathfrak{M}^2_{[0,t_m]}(\mathbf{G})$ of the previous mapping is called the *stochastic integral process of X with respect to M* and is denoted by $(\int X\,dM)$.

14.7 Properties of the Stochastic Integral

Proposition *We consider $M\in\mathfrak{M}^2_{[0,t_m]}(\mathbf{H})$, $X\in\Lambda^2(\mathbf{H};\mathbf{G};\mathcal{P},M)$, $N = (\int X\,dM)\in\mathfrak{M}^2(\mathbf{G})$. Then the following formulas hold:*

$$\alpha_N = \operatorname{Tr}(X\circ\tilde{Q}_M\circ X^*)\alpha_M; \tag{14.7.1}$$

$$\tilde{Q}_N = (\operatorname{Tr}(X\circ\tilde{Q}_M\circ X^*))^{-1}X\circ\tilde{Q}_M\circ X^*; \tag{14.7.2}$$

$$\mu_N = (X\circ\tilde{Q}_M\circ X^*)\alpha_M = (X\otimes X)\,\mu_M; \tag{14.7.3}$$

$$\langle N\rangle_t = \int_{]0,t]} \operatorname{Tr}(X\circ\tilde{Q}_M\circ X^*)\,d\langle M\rangle; \tag{14.7.4}$$

$$\langle\!\langle N\rangle\!\rangle_t = \int_{]0,t]} (X\circ\tilde{Q}_M\circ X^*)\,d\langle\!\langle M\rangle\!\rangle. \tag{14.7.5}$$

Proof From the definition and the isometry property we derive for every predictable rectangle $F \times]s, t]$

$$E\left(1_F \|N_t - N_s\|_{\mathbf{G}}^2\right) = E\left(\left\| \int_{F \times]s,t]} X\, dM \right\|_{\mathbf{G}}^2\right)$$
$$= \int_{F \times]s,t]} \mathrm{Tr}(X \circ \tilde{Q}_M \circ X^*) d\alpha_M.$$

This expresses (14.7.1). Since the right-hand side of (14.7.4) is clearly a predictable process, (14.7.4) follows readily from (14.7.1).

For every $X \in \mathcal{E}$ it is readily verified that

$$E\left\{1_F (N_t - N_s)^{\otimes_2}\right\} = \int_{F \times]s,t]} X \circ \tilde{Q}_M \circ X^*\, d\alpha_M. \qquad (14.7.6)$$

If X is an element of $\Lambda^2(\mathbf{H}; \mathbf{G}; \mathcal{P}, M)$, it can be approximated by a sequence (X_n) in $\mathcal{E}(\mathcal{L}(\mathbf{H}; \mathbf{G}))$, and an immediate continuity argument shows that (14.7.6) holds for such an X and the corresponding integral process $N := (\int X\, dM)$. Since (14.7.6) can be written

$$\mu_N = X \circ \tilde{Q}_M \circ X^* \alpha_M,$$

we obtain (14.7.3). The equality (14.7.2) now follows from (14.7.1) and (14.7.3). From the definition of $\langle M \rangle$, and according to (14.7.3), the Doléans measure of $(\int_{]0,t]} X_s \otimes X_s\, d\langle M \rangle_s)_{t \in T}$ is μ_N. But since the last process is predictable, it is precisely $\langle N \rangle$.

14.8 A Representation Theorem

Let $M_i \in \mathcal{M}_T^2(\mathbf{R})$. The *stable subspace of* $\mathcal{M}_T^2(\mathbf{R})$, *generated by* $\{M_i : i \in I\}$, is the smallest closed subspace of $\mathcal{M}_T^2(\mathbf{R})$ containing the images by $X \rightsquigarrow (\int X\, dM_i)$ of $L^2(\Omega', \mathcal{P}, \alpha_M)$.

The following proposition gives a (trivially necessary) sufficient condition for an element $U \in \mathcal{M}_T^2(\mathbf{G})$ to admit a representation $U = (\int X\, dM)$, where $M \in \mathcal{M}_T^2(\mathbf{H})$ and $X \in \Lambda^2(\mathbf{H}; \mathbf{G}; \mathcal{P}, M)$. In the finite-dimensional case the following statement can be found in [Mey-8] and [Gal]:

Proposition *Let* $M \in \mathcal{M}_T^2(\mathbf{H})$ *and* $U \in \mathcal{M}_T^2(\mathbf{G})$ *with* $T = [0, t_m]$, $M_0 = U_0 = 0$. *We assume that for every* $g \in \mathbf{G}$ *the real martingale* $\langle g, U \rangle_{\mathbf{G}}$ *belongs to the stable subspace of* $\mathcal{M}_T^2(\mathbf{R})$ *generated by* $(\langle h, M \rangle_{\mathbf{H}}, h \in \mathbf{H})$. *Then there exists* $\psi \in \Lambda^2(\mathbf{H}; \mathbf{G}; \mathcal{P}, M)$ *such that* $U = (\int \psi\, dM)$.

Proof We need a preliminary lemma.

Lemma *Let* $O_{\mathbf{H}}$ *be the set of unitary linear mappings from* $\mathbf{H}$ *into itself (with the uniform norm topology) and* (h_i) *be an orthonormal basis of* $\mathbf{H}$. *Then there exists a mapping* F *from* $\mathcal{L}_s^1(\mathbf{H}; \mathbf{H})$[1] *into* $O_{\mathbf{H}}$, *which is measurable when we*

[1] $\mathcal{L}_s^1(\mathbf{H}; \mathbf{H})$ stands for the space of symmetric nuclear operators in $\mathbf{H}$.

consider on $\mathfrak{L}_s^1(\mathbf{H};\mathbf{H})$ *and* $O_{\mathbf{H}}$ *the σ-algebras of universally measurable sets (i.e., the σ-algebras that are the intersections of all possible completions of the Borel σ-algebra) and such that* $F(v)\circ v\circ F^*(v)$ *is diagonal in the basis* (h_i) *for every* $v \in \mathfrak{L}_s^1(\mathbf{H};\mathbf{H})$.

To prove this lemma we consider the set

$$\{(v,u): v\in\mathfrak{L}_s^1(\mathbf{H};\mathbf{H}),\ u\in O_{\mathbf{H}},\ u\circ v\circ u^* \text{ is diagonal in } (h_i)\}.$$

Since this set is the intersection of the denumerably many sets

$$\{(v,u): \langle u\circ v\circ u^* h_i, h_j\rangle_{\mathbf{H}}\} = 0, \qquad i,j\in\mathbf{N},$$

it is clearly closed in the product space $\mathfrak{L}_s(\mathbf{H};\mathbf{H})\times O_{\mathbf{H}}$. We may then use a classical section theorem which states that if E_1 and E_2 are two complete metric spaces and if R is a Borel set in $E_1\times E_2$, the projection of which on E_1 is E_1 itself, there exists a mapping $v\to F(v)$ from E_1 into E_2 such that $(v,F(v))\in R$ for all $v\in E_1$ and F is measurable for the universal σ-algebras (for a proof of this theorem, see for example [DeM, Chapter III]). The above preliminary lemma follows.

A first consequence of this lemma is the following: let u be a symmetric nuclear operator in $\mathbf{H}$. We call u^+ the operator with domain $(\mathfrak{D}(u^+) := u(\mathbf{H})$ that in a basis of eigenvectors (e_i) of u is defined through the formula

$$u^+(h) := \sum_{\lambda_n\neq 0}\frac{1}{\lambda_n}(he_n)e_n,$$

where λ_n is the eigenvalue corresponding to e_n. With this notation we clearly have

$$(F(u)\circ u\circ F^*(u))^+ = F(u)\circ u^+\circ F^*(u).$$

Moreover, as a consequence of the same lemma the eigenvalues

$$\lambda_i(u) = \langle F(u)\circ u\circ F^*(u)h_i, h_i\rangle_{\mathbf{H}}$$

are measurable functions of u. We therefore conclude that the processes $F(\tilde{Q}_M)$ and $\tilde{Q}_M^+$, where $\tilde{Q}_M$ is the process defined in Section 14.4, are measurable for the α_M-completion of $\mathcal{P}$, which will give a meaning to the subsequent integrals.

By definition, if we set

$$X(\omega,s) := F(\tilde{Q}_M(\omega,s)) \qquad \text{for every} \quad (\omega,s),$$

the process $X(\omega,s)\circ\tilde{Q}_M(\omega,s)\circ X^*(\omega,s)$ has a diagonal representation in the basis (h_i), and moreover,

$$\mathrm{Tr}(X\circ\tilde{Q}_M\circ X^*) = \mathrm{Tr}\,\tilde{Q}_M.$$

The process X then belongs to $\Lambda^2(\mathbf{H}; \mathbf{H}; \mathcal{P}, M)$, and according to the proposition in Section 14.7, if we set $N := (\int X\, dM)$, we have $\tilde{Q}_N = X \circ \tilde{Q}_M \circ X^*$, and therefore the real martingales $N^i := \langle N, e_i \rangle_{\mathbf{H}}$ are strongly orthogonal. Since $X^* \circ \tilde{Q}_N \circ X = \tilde{Q}_M$, we also obtain

$$X^* \in \Lambda^2(\mathbf{H}; \mathbf{H}; \mathcal{P}, N) \qquad \text{and} \qquad M = \int X^*\, dN.$$

This shows in particular that the stable set in $\mathfrak{M}_T^2$ generated by $\{\langle h, M \rangle, h \in \mathbf{H}\}$ is included in the Hilbert closure of the subspace generated by the orthogonal subspaces $\{(\int \phi\, dN^i)\colon \phi \in \Lambda^2(\mathbf{R}; \mathbf{R}; \mathcal{P}, N^i)\}$.

Let (g_n) be an orthonormal basis in $\mathbf{G}$. For every n we may therefore write

$$\langle U, g_n \rangle_{\mathbf{G}} = \sum_i \left(\int \phi_{n,i}\, dN^i \right) \qquad (\text{sum in } \mathfrak{M}_T^2), \tag{14.8.1}$$

with for every $t \in T = [0, t_m]$

$$\begin{aligned} +\infty > \sum_n E(\langle U_t, g_n \rangle) &= \sum_n \sum_i E \int_{]0,t]} |\phi_{n,i}|^2\, d\langle N^i \rangle \\ &= \sum_n \int_{]0,t] \times \Omega} \sum_i |\phi_{n,i}|^2 \tilde{Q}_N^{ii}\, d\alpha_N. \end{aligned} \tag{14.8.2}$$

If we set

$$\Phi(\omega, t)(h) := \sum_n \left(\sum_i \phi_{n,i}(\omega, t) \langle h, e_i \rangle_{\mathbf{H}} \right) g_n,$$

we remark that

$$\mathcal{D}\Phi(\omega, t) \supset \left\{ h : \sum_{n,i} |\phi_{n,i}(\omega, t)|^2 (\langle h, e_i \rangle_{\mathbf{H}})^2 < \infty \right\},$$

and because of (14.8.2)

$$\mathcal{D}\Phi(\omega, t) \supset \left\{ \sum_{n,i} \sqrt{\tilde{Q}_N^{ii}}\, \langle h, e_i \rangle_{\mathbf{H}} e_i \right\} = \tilde{Q}_N^{1/2}(\mathbf{H}), \qquad \alpha_N \text{ a.e.}$$

Moreover, since

$$\|\Phi(\omega, t) \tilde{Q}_N^{1/2}\|_2^2 = \sum_{n,i} |\phi_{n,i}|^2 \tilde{Q}_N^{ii},$$

the process Φ belongs to $\Lambda^2(\mathbf{H}; \mathbf{G}; \mathcal{O}, M)$ and (14.8.1) then implies

$$U = \left(\int \Phi\, dN \right).$$

We finally get the formula of the theorem by setting

$$\psi = \Phi \circ X.$$

EXTENSIONS AND COMMENTS

1 Let $Z_1 \in \mathcal{M}_T^2(\mathbf{H})$ and $Z_2 \in \mathcal{M}_T^2(\mathbf{G})$, **H** and **G** being separable Hilbert spaces. The reader can find in [Ouv] the following extension of the preceding representation lemma: there exists $\Phi \in \Lambda^2(\mathbf{H}; \mathbf{G}; \mathcal{P}, Z_1)$ uniquely defined up to α_M-equivalence such that the **G**-valued martingale $Z_2 - \int(\Phi\, dZ_1)$ is strongly orthogonal to Z_1.

2 Using the stochastic calculus for Hilbert-valued martingales, the following extension of a classical Girsanov formula may be proved: $M \in \mathcal{M}_{[0,1]}^2(\mathbf{H})$, Φ is an **H**-valued predictable process such that a.s. the mapping $t \rightsquigarrow \tilde{Q}_M(\cdot, t)\circ\Phi(\cdot, t)$ is integrable on [0, 1]. Call Y the process such that

$$Y_t := \int_{]0,t]} \tilde{Q}_M(\cdot, s)\circ\Phi(\cdot, s)\, d\langle M\rangle_s + M_t,$$

assume that

$$E\left(\int_{[0,1]} \langle \Phi_s, \tilde{Q}_M \circ \Phi_s\rangle_{\mathbf{H}}\, d\langle M\rangle_s\right) < \infty.$$

We set

$$Z_t := \exp\left\{ -\int_{]0,t]} \Phi_s^*\, dM_s - \frac{1}{2}\int_{]0,t]} \langle \Phi_s^*, \tilde{Q}_M(s)\circ\Phi(s)\rangle_{\mathbf{H}}\, d\langle M\rangle_s \right\}$$

and assume that

$$E(Z_1) = 1.$$

Then the process Y is a square integrable martingale for the probability $Z_1 P$ and, moreover, for this probability $\langle Y\rangle = \langle M\rangle$.

The proof consists in showing, by a simple calculation involving the exponential formula of Exercise 3 in Section 4.12, that $(Z_t U_t^h)_{t\in[0,1]}$ is a P-martingale when we set

$$U_n^h := \exp\left\{ \langle h, Y_t\rangle_{\mathbf{H}} - \frac{1}{2}\int_{]0,t]} \langle h, \tilde{Q}_M(s)(h)\rangle_{\mathbf{H}}\, d\langle M\rangle_s \right\}.$$

15 CYLINDRICAL PROCESSES

As mentioned in Section 4, a Hilbert-valued process with independent increments has a "covariance" which is an element of $\mathbf{H}\hat{\otimes}_1\mathbf{H}$. It cannot be a bilinear form associated, for example, with the "unit matrix" of **H**, or in terms of the identifications with linear operators made in Section 14.2, the covariance of a Hilbert-valued process cannot be the identity map of **H** onto **H** when **H** is infinite dimensional. Cylindrical processes have been introduced to give, for example, a meaning to Hilbert-valued brownian motion, the covariance of which is the identity operator (see, for example, [Ben] and [Par-2] for the use of such a concept to describe random perturbations).

15.1 Cylindrical Random Elements

B is a separable Banach space with dual **B**′. A *p-cylindrical* **B**-*random element* is by definition a continuous linear mapping from **B** into $L^p(\Omega, \mathfrak{F}, P)$. This is a notion equivalent to the notion of the generalized process in [GeW]. We speak of cylindrical random elements in reference to the notion of cylindrical measure as treated, for example, in [Bad].

The *covariance* of a 2-cylindrical **B**-random element $\tilde{U}$ is the bilinear form $(h, g) \rightsquigarrow C_{\tilde{U}}(h, g) := E(\tilde{U}(h)\tilde{U}(g))$. If U is a **B**′-valued random variable in the ordinary sense, the mapping $h \rightsquigarrow \langle h, U \rangle$ defines a p-cylindrical **B**-random element iff $E(\|U\|^p) < \infty$.

A question is, Conversely, given a p-cylindrical **B**-random element $\tilde{U}$, does there exist an ordinary sense **B**-valued random variable U such that $\tilde{U}(h) = \langle h, U \rangle$? We do not want to give a general answer to this difficult problem. The reader is referred to [BaC]. A particular answer in the case of Hilbert spaces is easy. It is the following.

Theorem *If* **H** *is a separable Hilbert space, a 2-cylindrical random element $\tilde{U}$ is associated with an ordinary sense* **H**-*valued random variable U if and only if $\tilde{U}$ is a Hilbert–Schmidt operator from* **H** *into $L^2(\Omega, \mathfrak{F}, P)$. We then have $\|\tilde{U}\|_2^2 = E\|U\|_{\mathbf{H}}^2$.*

Proof Let (h_n) be an orthonormal basis in **H**. If U is a square integrable **H**-valued random variable, then

$$\sum_n E|\langle h_n, U \rangle|^2 = E\|U\|^2 < \infty.$$

This proves that the mapping $h \rightsquigarrow \langle h, U \rangle$ is Hilbert–Schmidt (cf. Section 14.2(1)) with a Hilbert–Schmidt norm equal to $E\|U\|^2$. Conversely, if $\tilde{U} \in \mathfrak{L}_2(\mathbf{H}; L^2(\Omega, \mathfrak{F}, P))$, we may set

$$U := \sum_n \tilde{U}(h_n)h_n,$$

and this defines (taking for each n a particular member of the class $\tilde{U}(h_n)$ of real random variables) an **H**-valued square integrable random variable because of the Hilbert–Schmidt property of $\tilde{U}$, which ensures that

$$\sum_n E|\tilde{U}(h_n)|^2 < \infty.$$

In the case of Banach spaces the situation is far from being as clear as that for Hilbert spaces (see, for example, [Bad], [BaC], and [Che]).

15.2 Cylindrical Processes

In order to simplify the notation, we shall write $\mathfrak{L}_t^p$ for the space $\mathfrak{L}^p(\Omega, \mathfrak{F}_t, P)$ and L_t^p for $L^p(\Omega, \mathfrak{F}_t, P)$.

Definitions Let $p > 0$. We call *p-cylindrical* **B**-*process*, where **B** is a Banach space, any family $\tilde{X} := \{X_t : t \in T,\ \tilde{X}_t \in \mathfrak{L}(\mathbf{B}; L_t^p)\}$. The p-cylindrical **B**-process $\tilde{X}$ is said to be *weakly cadlag* if for every $h \in \mathbf{B}$ there is a cadlag version of the adapted real process $(\tilde{X}_t)_{t \in T}$.

If $p \geq 1$ and if the real process $\{\tilde{X}_t(h) : t \in T\}$ is a martingale for every $h \in \mathbf{B}$, $\tilde{X}$ will be called a *p-cylindrical martingale*.

The Doléans function $d_{\tilde{X}}$ of a 1-cylindrical **B**-process $\tilde{X}$ is the additive **B**′-valued function on $\mathcal{Q}$ defined by

$$\langle h, d_{\tilde{X}}(F \times]s,t])\rangle := E\left[1_F(\tilde{X}_t(h) - \tilde{X}_s(h))\right]$$

for every predictable rectangle $F \times]s,t]$. If this function has a σ-additive extension to $\mathcal{P}$, this extension is called the Doléans measure of $\tilde{X}$.

When X is an ordinary sense **B**′-valued process such that $E\|X\|^p < \infty$, the family $(\tilde{X}_t)_{t \in T}$ of mappings $h \mapsto \langle h, \tilde{X}_t\rangle$ defines a p-cylindrical **B**-process.

It will immediately be seen by the reader that d_X and $d_{\tilde{X}}$ are related by $\langle h, d_X\rangle = d_{\tilde{X}}(h)$.

Theorem *Let $\tilde{X}$ be a 1-cylindrical* **B**-*process such that for every $h \in \mathbf{B}$ the Doléans measure $d_{\tilde{X}}^h$ of the real process $\tilde{X}(h)$ exists and such that $\sup_{A \in \mathcal{Q}} \|d_{\tilde{X}}(A)\|_{\mathbf{B}'} < \infty$. Then $d_{\tilde{X}}$ has a σ-additive* **B**′-*valued extension to $\mathcal{P}$.*

Proof This theorem could be considered as an immediate consequence of a general theorem for vector measures. We give a particular proof here for the sake of completeness.

We call **B*** the algebraic dual of **B** (the space of all linear forms on **B**). We define immediately a **B***-valued extension of $d_{\tilde{X}}$ by setting

$$\forall G \in \mathcal{P}, \quad h \in \mathbf{B}, \qquad \langle h, d_{\tilde{X}}(G)\rangle := d_{\tilde{X}}^h(G).$$

We have to prove that this extension actually takes its values in $\mathbf{B}' \subset \mathbf{B}^*$. We consider on **B*** the topology usually called $\sigma(\mathbf{B}^*, \mathbf{B})$, which is the topology of pointwise convergence on **B**. This topology induces the topology $\sigma(\mathbf{B}', \mathbf{B})$ on **B**′. The classical Alaoglu theorem (see [DuS]) states that the bounded sets in **B**′, closed for $\sigma(\mathbf{B}', \mathbf{B})$, are compact for this topology. The closed convex hull in **B*** of a set in **B**′ for the $\sigma(\mathbf{B}^*, \mathbf{B})$ topology is therefore a set in **B**′ (see [DuS, Chapter V]). But for every $h \in \mathbf{B}$ and $r \in \mathbf{R}$, it is clear that the inequality $\inf_{A \in \mathcal{Q}} \langle h, d_{\tilde{X}}(A)\rangle \geq r$ implies $\inf_{\mathbf{G} \in \mathcal{P}} d_{\tilde{X}}^h(G) \geq r$. This expresses that $\{d_{\tilde{X}}(G) : G \in \mathcal{P}\}$ is included in the closed convex hull in **B*** of the set $\{d_{\tilde{X}}(A) : A \in \mathcal{Q}\} \subset \mathbf{B}'$. Therefore $d_{\tilde{X}}(G) \in \mathbf{B}'$ for all $G \in \mathcal{P}$ because of the boundedness assumption on $\{d_{\tilde{X}}(A) : A \in \mathcal{Q}\}$ and following the previous argument.

By construction the mapping $G \rightarrow d_{\tilde{X}}(G)$ is σ-additive for the topology $\sigma(\mathbf{B}', \mathbf{B})$. But this implies (see [DuS, Chapter IV]) that it is also σ-additive for the norm of $\mathbf{B}'$. This proves the theorem.

15.3 The Quadratic Doléans Measure of a 2-Cylindrical Martingale

Let $\tilde{M}$ be a 2-cylindrical $\mathbf{B}$-martingale. Let us call (rather improperly) $\tilde{M}_t \otimes \tilde{M}_t$ the continuous linear mapping from $\mathbf{B} \hat{\otimes}_1 \mathbf{B}$ into L_t^p, which is the linear continuous extension of $h \otimes g \rightarrow \tilde{M}_t(h)\tilde{M}_t(g)$.

If $\tilde{M}$ is associated with the ordinary sense square integrable $\mathbf{B}'$-valued martingale M, the following equality holds:

$$\langle M_t \otimes M_t, h \otimes g \rangle = \langle M_t, h \rangle \langle M_t, g \rangle.$$

The natural extension of the concept of the Doléans measure of $M \otimes M$ is the Doléans measure of $\tilde{M} \otimes \tilde{M}$ as just defined. This measure, if it exists, will be called the *quadratic Doléans measure of* $\tilde{M}$.

Proposition *Let $\tilde{M}$ be a 2-cylindrical* $\mathbf{B}$*-martingale on $[0, t_m]$ such that for every h, $\tilde{M}(h)$ has a cadlag version. Then it has a quadratic Doléans measure.*

Proof Because of the right continuity of the real process $\tilde{M}(h)$, the first assumption of the theorem of Section 15.2 is fulfilled. The boundedness property of $d_{M \otimes M}$ follows from

$$\begin{aligned} E\{1_F \langle \tilde{M}_t \otimes \tilde{M}_t - \tilde{M}_s \otimes \tilde{M}_s, h \otimes g \rangle\} &= E[1_F(\tilde{M}_t - \tilde{M}_s)(h)(\tilde{M}_t - \tilde{M}_s)(g)] \\ &\leqslant \sqrt{E\{1_F[(\tilde{M}_t - \tilde{M}_s)(h)]^2\}} \\ &\quad \times \sqrt{E\{1_F[(\tilde{M}_t - \tilde{M}_s)(g)]^2\}} \\ &\leqslant \sqrt{E[\tilde{M}_{t_m}(h)]^2} \sqrt{E[\tilde{M}_{t_m}(g)]^2} \\ &\leqslant \|\tilde{M}_{t_m}\| \, \|h\| \, \|g\| \end{aligned}$$

for every predictable rectangle $F \times]s, t]$. An analogous argument can be used for any $A = \sum_i F_i \times]s_i, t_i]$ by noticing that

$$\sum_i E[1_{F_i}(\tilde{M}_{t_i} - \tilde{M}_{s_i})(h)]^2 \leqslant E[|\tilde{M}_{t_m}(h)|^2].$$

This proves the proposition.

15.4 Cylindrical Standard Brownian Motion

As already noticed, there exists no **B**′-valued process X with independent increments such that the covariance of $X_t - X_s$ is tC, where C is the identity mapping in **B**′.

Now if $\tilde{W}$ is a 2-cyclindrical **B**′-process such that for every $h \in \mathbf{B}$ the real process $\tilde{W}(h)$ is a brownian motion, the quadratic Doléans measure of $\tilde{W}$ exists, according to Section 15.3, and is a $(\mathbf{B} \hat{\otimes}_1 \mathbf{B})'$-valued measure that is the extension of $F \times]s, t] \rightsquigarrow (t-s)P(F)C$, where C is the bilinear form

$$tC(h, g) = E(\tilde{M}_t(h)\tilde{M}_t(g)).$$

In other words the quadratic Doléans measure is $(P \otimes l)C$, where l is the Lebesgue measure and C a bilinear form.

We leave it to the reader to prove, by using a standard argument with gaussian processes, that given any bilinear form C, a cylindrical process $\tilde{W}$ exists such that $\tilde{W}(h)$ is a brownian motion for every $h \in \mathbf{B}$ and such that $E((\tilde{M}_t(h))(\tilde{M}_t(g)) = C(h, g)$ for every $(h, g) \in \mathbf{B} \times \mathbf{B}$. Such a cylindrical process will be called a *cylindrical brownian motion*.

If instead of a general separable Banach space **B** we consider a separable Hilbert space **H**, a 2-cylindrical **H**-brownian motion is said to be *standard* if $C(h, g) = hg$ (in other words C is associated with the identity mapping of **H**).

15.5 The Isometry $\mathfrak{M}_T^2(\mathbf{H}) \approx \mathfrak{L}_2(\mathbf{H}; \mathfrak{M}_T^2)$

Let **H** be a separable Hilbert space. Referring to the theorem in Section 15.1, we see that a 2-cylindrical **H**-martingale $\tilde{M}$ is associated with $M \in \mathfrak{M}_T^2(\mathbf{H})$ iff the mapping $h \rightsquigarrow \tilde{M}_t(h)$ belongs to $\mathfrak{L}_2(\mathbf{H}; L^2(\Omega, \mathfrak{F}, P))$ for every $t \in [0, t_m]$.

If (h_n) is an orthonormal basis in **H**, we have

$$\sum_n \|\tilde{M}(h_n)\|^2_{\mathfrak{M}_T^2} = \sum_n E|\tilde{M}_{t_m}(h_n)|^2 = \|\tilde{M}_{t_m}\|_2^2 = E\|M_{t_m}\|^2_{\mathbf{H}}.$$

This precisely expresses that the correspondance $M \rightarrow \tilde{M}$ is an isometry between $\mathfrak{L}_2(\mathbf{H}; \mathfrak{M}_T^2)$ and $\mathfrak{M}_T^2(\mathbf{H})$.

15.6 Meyer Processes and Connected Comments

1 The considerations of Section 9 can be applied immediately to $(\mathbf{B} \hat{\otimes}_1 \mathbf{B})'$-valued admissible measures as soon as they are of bounded variation. In particular, we may speak of the Meyer process $\langle \tilde{M} \rangle$ as being the Meyer process of the $(\mathbf{B} \hat{\otimes}_1 \mathbf{B})'$-valued quadratic Doléans measure of $\tilde{M}$ when this one has bounded variation. In the case of a cylindrical brownian motion $\tilde{N}$ with covariance C as defined above, we clearly have $\langle \tilde{W} \rangle_t = tC$.

2 It should be noticed that contrary to the case of ordinary sense square integrable martingales, the quadratic Doléans measure is not always of bounded variation. We suggest a counterexample to the reader:

Let (t_n) be a decreasing sequence of real numbers, vanishing to zero. Let **H** be a separable Hilbert space and (e_n) some orthogonal basis of **H**. Show that the 2-cylindrical martingale $\tilde{M}$, defined by

$$\tilde{M}_t(h) := \sum_n (he_n)\left(\sqrt{t_n - t_{n+1}}\right)^{-1}\left(\beta_{t_n \wedge t} - \beta_{t_{n+1} \wedge t}\right),$$

where β is a real standard brownian motion, has a quadratic Doléans measure of unbounded variation (for the partition $\{\Omega \times]t_{n+1}, t_n]\}$ of Ω',

$$\sum_n \|d_{\tilde{M}\otimes\tilde{M}}(\Omega \times]t_{n+1}, t_n])\| \geqslant \sum_n \frac{1}{t_n - t_{n+1}} E\left(\beta_{t_n} - \beta_{t_{n+1}}\right)^2 = +\infty.)$$

16 STOCHASTIC INTEGRAL WITH RESPECT TO 2-CYLINDRICAL MARTINGALES WITH FINITE QUADRATIC VARIATION

16.1 2-Cylindrical Martingales with Finite Quadratic Variation

A 2-cylindrical martingale $\tilde{M}$ will be said to have finite quadratic variation if the quadratic Doléans measure associated with $\tilde{M}$ (see Section 15.3) has bounded variation. The essential difference from the case in which M is an ordinary sense process lies in the fact that $(\mathbf{B}\hat{\otimes}_1\mathbf{B})'$, the space in which $d_{\tilde{M}\otimes\tilde{M}}$ takes its values, is not in general a Banach space with the Radon–Nikodym property.

Let us call $\mu_{\tilde{M}}$ the quadratic Doléans measure of $\tilde{M}$ and $|\mu_{\tilde{M}}|$ its variation. We cannot speak of the predictable density of $\mu_{\tilde{M}}$ with respect to $|\mu_{\tilde{M}}|$ as we did in Section 14 for μ_M and $|\mu_M|$.

For the first time in this book we have to consider weakly measurable vector valued functions.

Proposition *Let $\tilde{M}$ be a 2-cylindrical* **B**-*martingale, $\mu_{\tilde{M}}$ its quadratic Doléans measure, and $|\mu_{\tilde{M}}|$ the variation of this measure, assumed to be bounded. There is a process $Q_{\tilde{M}}$ with values in the set of positive elements of $(\mathbf{B}\hat{\otimes}_1\mathbf{B})'$ such that for every $\xi \in (\mathbf{B}\hat{\otimes}_1\mathbf{B})$ the real process $\langle \xi, Q_{\tilde{M}}\rangle$ is measurable for the $|\mu_{\tilde{M}}|$-completion of $\mathcal{P}$, is defined uniquely up to $|\mu_{\tilde{M}}|$-equivalence, and has the property*

$$\langle \xi, \mu_{\tilde{M}}(A)\rangle = \int_A \langle \xi, Q_{\tilde{M}}(\omega, s)\rangle |\mu_{\tilde{M}}|(d\omega, ds) \qquad \forall \xi \in (\mathbf{B}\hat{\otimes}_1\mathbf{B}), \quad A \in \mathcal{P} \tag{16.1.1}$$

and

$$\|Q_{\tilde{M}}\| = 1, \qquad |\mu_{\tilde{M}}| \text{ a.e.}$$

Proof This proposition may be considered to be an immediate consequence of a weak Radon–Nikodym theorem as, for example, in [Met-7, Part II]. To be more self-contained we sketch a direct proof, the details of which will easily be supplied by the reader. For every $\xi \in (\mathbf{B} \hat{\otimes}_1 \mathbf{B})$ take the Radon–Nikodym density of the real measure $\langle \xi, \mu_M \rangle$ with respect to $|\mu_M|$. Observe that this density $q^{\xi}(\omega, s)$ is bounded by $\|\xi\|$ and that the mapping $\xi \rightsquigarrow q^{\xi}(\cdot)$ is continuous from $\mathbf{B} \hat{\otimes}_1 \mathbf{B}$ into $L^{\infty}(\Omega', \mathcal{P}, |\mu|_M)$. We can therefore take a "lifting,". i.e., for every ξ choose a representative of the element q^{ξ} of $L^{\infty}(\Omega', \mathcal{P}, |\mu|_M)$ such that for every (ω, s) the mapping $\xi \rightsquigarrow q^{\xi}(\omega, s)$ for every (ω, s) is linear, with $q^{\xi}(\xi, s) \leqslant \|\xi\|$ (for the existence of such a "lifting" see [Iot-2]). If we write $Q_{\tilde{M}}(\omega, s)$ for the bounded linear form $\xi \rightsquigarrow q^{\xi}(\omega, s)$, we get a function $Q_{\tilde{M}}$ satisfying (16.1.1) and $\|Q_{\tilde{M}}\| \leqslant 1 |\mu_M|$ a.e. Since $\|Q_{\tilde{M}}(\omega, t)\| = \sup_{\xi \in \mathbf{D}} |\langle \xi, Q_{\tilde{M}}(\omega, t) \rangle|$ for some denumerable dense subset $\mathbf{D}$ of $\mathbf{B} \hat{\otimes}_1 \mathbf{B}$, the predictability of $Q_{\tilde{M}}$ is clear. Should the inequality $\|Q_{\tilde{M}}\| \leqslant \alpha < 1$ hold on some set A with $A \in \mathcal{P}$ and $|\mu_M|(A) > 0$, one could then easily derive a contradiction with the definition of $|\mu_M|$. Therefore $\|Q_{\tilde{M}}\| = 1$ holds $|\mu_{\tilde{M}}|$ a.e.

16.2 Stochastic Integral of Simple Processes

Since cylindrical $\mathbf{B}$-processes are a generalization of ordinary sense $\mathbf{B}'$-valued processes, the simple processes to start with are clearly the $\mathcal{L}(\mathbf{B}', \mathbf{K}')$-valued ones, $\mathbf{B}$ and $\mathbf{K}$ being Banach spaces.

Let us then take $\mathbf{L} := \mathcal{L}(\mathbf{B}'; \mathbf{K}')$ and consider the space $\mathcal{E}(\mathbf{L})$ as defined in Section 2. Let $Y \in \mathcal{E}(\mathbf{L})$ with $Y := \sum_{i \in I} u_i 1_{F_i \times]s_i, t_i]}$, $u_i \in \mathbf{L}$, I finite, and $F_i \times]s_i, t_i]$ a predictable rectangle.

We assume that $\mathbf{B}$ is *reflexive*, and for every $u \in \mathcal{L}(\mathbf{B}'; \mathbf{K}')$ we call u^* the transposed mapping from $\mathbf{K}$ into $\mathbf{B}$, defined by

$$\langle u^*(k), b' \rangle = \langle k, u(b') \rangle \qquad \text{for every} \quad k \in \mathbf{K}, \quad b' \in \mathbf{B}'.$$

For every $k \in \mathbf{K}$ the process $\tilde{M}(u_i^*(k))$ is by definition a real square integrable martingale with norm in $\mathfrak{M}_T^2$ smaller than $\|u_i^*\| \, \|k\|$ ($T := [0, t_m]$).

By definition we call stochastic integral process of Y with respect to $\tilde{M}$ the 2-cylindrical $\mathbf{K}$-martingale $\tilde{Z}$ defined by

$$(\tilde{Z}(k))_t := \sum_{i \in I} 1_{F_i} \{ (\tilde{M}(u_i^*(k)))_{t_i \wedge t} - (\tilde{M}(u_i^*(k)))_{s_i \wedge t} \} \qquad \forall k \in \mathbf{K} \tag{16.2.1}$$

and denoted by $(\int Y\,d\tilde{M})$. For every $k \in \mathbf{K}$ the norm in $\mathfrak{M}_T^2$ of the square integrable martingale $(\int Y\,d\tilde{M})(k)$ is given by

$$\left\|\left(\int Y\,d\tilde{M}\right)(k)\right\|_{\mathfrak{M}_T}^2 = \sum_{i \in I} E\{1_{F_i}(\tilde{M}_{t_i} - \tilde{M}_{s_i})(u_i^*(k))\}$$
$$= \sum_{i \in I} \langle u_i^*(k) \otimes u_i^*(k),\ \mu_{\tilde{M}}(F_i \times]s_i, t_i])\rangle.$$

Following the definition of $Q_{\tilde{M}}$, this equality can be written

$$\left\|\left(\int Y\,d\tilde{M}\right)(k)\right\|_{\mathfrak{M}_T}^2 = \int \langle Y^*(k) \otimes Y^*(k), Q_{\tilde{M}}\rangle\, d|\mu_{\tilde{M}}|. \qquad (16.2.2)$$

16.3 Integrable Processes for the Isometric Integral

Formula (16.2.2) suggests the introduction of the following space endowed with a pre-Hilbert structure.

Definition $\tilde{L}(\tilde{M}; \mathbf{B}'; \mathbf{K}')$ is the set of processes X with the following properties:

(i) for every $(\omega, t) \in \Omega \times T$, $X(\omega, t)$ is a linear operator from $\mathbf{B}'$ into $\mathbf{K}'$ with domain $\mathfrak{D}_X(\omega, t)$ dense in $\mathbf{B}'$;

(ii) denoting by $X^*(\omega, t)$ the adjoint of $X(\omega, t)$, which exists as an operator from $\mathbf{K}$ into $\mathbf{B}$, in view of (i) and the reflexivity of $\mathbf{B}$, the bilinear form $(k_1, k_2) \rightsquigarrow \langle X^*(\omega, t)(k_1) \otimes X^*(\omega, t)(k_2), Q_{\tilde{M}}(\omega, t)\rangle$ has $|\mu_{\tilde{M}}|$-a.e. a unique continuous extension to $\mathbf{K} \times \mathbf{K}$, the process $\langle X^*(k_1) \otimes X^*(k_2), Q_{\tilde{M}}\rangle$ being predictable for every $(k_1, k_2) \in \mathbf{K} \times \mathbf{K}$;

(iii) $N(X) := \sup_{\|k\| \leq 1}[\int_{\Omega'} \langle X^*(k) \otimes X^*(k), Q_{\tilde{M}}\rangle\, d(|\mu_{\tilde{M}}|)]^{1/2} < \infty$.

It is clear that N is a seminorm on $\tilde{L}(\tilde{M}; \mathbf{B}'; \mathbf{K}')$, and we call $\tilde{\Lambda}(\tilde{M}; \mathbf{B}'; \mathbf{K}')$ the closure of $\mathcal{E}(\mathfrak{L}(\mathbf{B}'; \mathbf{K}'))$ in this space. Unfortunately, as shown by the example in Section 16.6 below, the space $\tilde{\Lambda}(\tilde{M}; \mathbf{B}'; \mathbf{K}')$ *is not complete.*

Proposition *The space $L^2_{\mathfrak{L}(\mathbf{B}'; \mathbf{K}')}(\Omega', \mathcal{P}, |\mu_{\tilde{M}}|)$, abbreviated as $L^2(\tilde{M}; \mathbf{B}'; \mathbf{K}')$, is included in $\tilde{\Lambda}(\tilde{M}; \mathbf{B}'; \mathbf{K}')$, and the L^2-norm $\|X\|_2$ of $X \in L^2(\tilde{M}; \mathbf{B}'; \mathbf{K}')$ satisfies $N(X) \leq \|X\|_2$.*

Proof If $X \in L^2(\tilde{M}; \mathbf{B}'; \mathbf{K}')$, we immediately get

$$N(X) \leq \sup_{\|k\| \leq 1} \left[\int_{\Omega'} \|X^*(k)\|_{\mathbf{B}}^2 \|Q_{\tilde{M}}\|^2\, d|\mu_{\tilde{M}}|\right]^{1/2} \leq \left[\int_{\Omega'} \|X^*\|_{\mathbf{B}}^2\, d|\mu_{\tilde{M}}|\right]^{1/2}.$$

Since $L^2(\tilde{M}; \mathbf{B}'; \mathbf{K}')$ lies in the closure of $\mathcal{E}(\mathfrak{L}(\mathbf{B}'; \mathbf{K}'))$ for the L^2-norm, the proposition is proved.

As an immediate consequence of (16.2.2), we see that *the mapping* $X \rightsquigarrow (\int X\, d\tilde{M})$ *from* $\mathcal{E}(\mathcal{L}(\mathbf{B}'; \mathbf{K}'))$ *into the space* $\tilde{\mathcal{L}}(\mathbf{K}'; \mathcal{M}_T^2)$ *of 2-cylindrical martingales has a unique isometric extension to* $\tilde{\Lambda}(\tilde{M}; \mathbf{B}'; \mathbf{K}')$. This extension is still noted $X \rightsquigarrow (\int X\, d\tilde{M})$, and the 2-cylindrical martingale $(\int Y\, d\tilde{M})$ is called the *stochastic integral of X*.

16.4. Application: An Interpretation of an Example of Yor

We have already mentioned in Section 2 the difficulties that arise when dealing with operator-valued processes, taking their values in $\mathcal{L}(\mathbf{B}; \mathbf{K})$, when $\mathbf{K}$ is a Banach but not a Hilbert space. Even when $\mathbf{B}$ is the one-dimensional real vector space, a bounded predictable process may admit no ordinary sense integral process (see Yor's example in Section 2.1 or [Yor-1]).

We come back to this example for which $\tilde{M}$ is the usual standard one-dimensional brownian motion β and (t_n) is a decreasing sequence of times chosen in such a way that

$$|\beta_{t_n} - \beta_{t_{n+1}}| = +\infty \text{ a.s.}$$

The process X under consideration is deterministic, with values in $l^1 \subset (l^\infty)'$ defined by $X(t) := (1_{]t_{n+1}, t_n[}(t))_{n \in \mathbf{N}}$. *Since the sequence* $(\beta_{t_n}(\omega) - \beta_{t_{n+1}}(\omega))_{n \in \mathbf{N}}$ is almost surely not an element of l^1, there is no ordinary sense stochastic integral process $(\int Y\, d\beta)$. But since Y is predictable and bounded as a process with values in $l^1 \subset (l^\infty)'$, we have a cylindrical integral process which actually is

$$\left(\int Y\, d\beta\right)_t (k) = \sum_n k_n(\beta_{t_n \wedge t} - \beta_{t_{n+1} \wedge t}) \qquad \text{for every} \quad k := (k_n)_{n \in \mathbf{N}} \in l^\infty,$$

the above series converging in $L^2(\Omega, \mathcal{F}_t, P)$.

16.5 The Radonifying Integral

The notion of the radonifying integral was introduced by B. Gaveau (see [Gav]) in order to integrate with respect to a standard infinite-dimensional brownian process. This extends naturally to 2-cylindrical $\mathbf{H}$-martingales, where $\mathbf{H}$ is a Hilbert space.

We now consider a separable Hilbert space $\mathbf{H}$ instead of a separable reflexive Banach space $\mathbf{B}$; $\mathbf{G}$ will denote another separable Hilbert space. As usual we identify $\mathbf{H}$ and $\mathbf{G}$ with their duals.

The term *radonifying* comes from the theory of measures in vector spaces (see [BaC], [Bad], for example): The "probability law" of a cylindrical random element is a cylindrical measure, while the probability law of an ordinary sense random vector is a measure with sufficiently many regularity properties to have been called a Radon measure (see [Sch]). The integral of a process X with respect to a cylindrical process is called radonifying when the integral cylindrical process is the one associated with an ordinary sense process.

Considering the $\mathcal{L}(\mathbf{H};\mathbf{H})$-valued process $\tilde{Q}_{\tilde{M}}$ defined by

$$\tilde{Q}_{\tilde{M}}(h)h' = Q_{\tilde{M}}(h\otimes h') \qquad \text{for every} \quad (h,h')\in \mathbf{H}\times\mathbf{H},$$

we may write for a process X in $\tilde{L}(\tilde{M};\mathbf{H};\mathbf{G})$

$$\begin{aligned}\langle X^*(g_1)\circ X^*(g_2), Q_M\rangle &= \langle \tilde{Q}_{\tilde{M}}\circ X^*(g_1), X^*(g_2)\rangle_{\mathbf{H}}\\ &= \langle X\circ\tilde{Q}_{\tilde{M}}\circ X^*(g_1), g_2\rangle_{\mathbf{G}}.\end{aligned}$$

Therefore we obtain trivially the inequality

$$(N(X))^2 \leqslant \sup_{\|g\|\leqslant 1}\int_{\Omega'}\langle X\circ\tilde{Q}_{\tilde{M}}\circ X^*(g), g\rangle\, d|\mu_{\tilde{M}}|. \tag{16.5.1}$$

Let us remark that the trace of the operator $u\circ\tilde{Q}_{\tilde{M}}\circ X^*$ is given by

$$\operatorname{Tr}(u\circ\tilde{Q}_{\tilde{M}}\circ X^*) = \sum_n \langle u\circ\tilde{Q}_{\tilde{M}}\circ X^*(g_n), g_n\rangle$$

for every orthonormal basis (g_n) in $\mathbf{G}$, and that the operator $u\circ\tilde{Q}_{\tilde{M}}\circ u^*$ is nuclear iff its trace is finite (see, for example, [GeW]). For every process X in $\tilde{\Lambda}(\tilde{M};\mathbf{H};\mathbf{G})$, the $\overline{\mathbf{R}}^+$-valued process $\operatorname{Tr}(X\circ\tilde{Q}_{\tilde{M}}\circ X^*)$ is predictable.

We call $\tilde{\Lambda}^2(\tilde{M};\mathbf{H};\mathbf{G})$ the subspace of $\tilde{\Lambda}(\tilde{M};\mathbf{H};\mathbf{G})$ consisting of processes X such that

$$\int_{\Omega'}\operatorname{Tr}(X\circ\tilde{Q}_{\tilde{M}}\circ X^*)\,d|\mu_{\tilde{M}}| < \infty. \tag{16.5.2}$$

This space clearly contains the processes $\mathcal{E}(\mathcal{L}_2(\mathbf{H};\mathbf{G}))$ and more generally the predictable processes X with values in $\mathcal{L}(\mathbf{H};\mathbf{G})$ such that

$$\int_{\Omega'}\|X\|_2^2\,d|\mu_{\tilde{M}}| < \infty.$$

We then have the following result.

Theorem (1) *The positive bilinear form* $(X,Y)\rightsquigarrow\int\operatorname{Tr}(X\circ\tilde{Q}_{\tilde{M}}\circ Y^*)\,d|\mu_{\tilde{M}}|$ *induces a hilbert structure on* $\tilde{\Lambda}^2(\tilde{M};\mathbf{H};\mathbf{G})$.

(2) *The mapping* $X\rightsquigarrow(\int X\,d\tilde{M})$ *is an isometry from* $\tilde{\Lambda}^2(\tilde{M};\mathbf{H};\mathbf{G})$ *into* $\Lambda^2(\mathbf{G};\mathfrak{M}_T^2)\approx\mathfrak{M}_T^2(\mathbf{G})$.

Proof (1) From the symmetry and positivity of $\tilde{Q}_{\tilde{M}}$ we may deduce the existence of $\tilde{Q}_{\tilde{M}}^{1/2}$, the operators $\tilde{Q}_{\tilde{M}}^{1/2}(\omega,t)$ being positive symmetric and such that $\tilde{Q}_{\tilde{M}} = \tilde{Q}_{\tilde{M}}^{1/2}\circ\tilde{Q}_{\tilde{M}}^{1/2}$ (see [Rud, Chapter 12]). The operators $X\circ\tilde{Q}_{\tilde{M}}^{1/2}(\omega,t)$, the squares of which are nuclear, are then Hilbert–Schmidt operators, and one may write

$$\operatorname{Tr}(X \circ \tilde{Q}_{\tilde{M}}^{1/2} \circ X^*) = \|X \circ \tilde{Q}_{\tilde{M}}^{1/2}\|_2^2.$$

We can use the same argument as that used in the proof of the proposition of Section 14.4 to prove the first part of the theorem.

(2) By considering any orthonormal basis (g_n) in $\mathbf{G}$ and applying formula (16.2.2), we get for $(\int X\, d\tilde{M})$ considered as a linear operator from $\mathbf{G}$ into $\mathfrak{M}_T^2$

$$\left\|\left(\int X\, d\tilde{M}\right)\right\|_2^2 = \sum_n \left\|\left(\int X\, d\tilde{M}\right)(g_n)\right\|_2^2 = \int_{\Omega'} \sum_n \langle X \circ \tilde{Q}_{\tilde{M}} \circ X^*(g_n), g_n\rangle_{\mathbf{G}}\, d|\mu_{\tilde{M}}|$$

$$= \int_{\Omega'} \operatorname{Tr}(X \circ \tilde{Q}_{\tilde{M}} \circ X^*)\, d|\mu_{\tilde{M}}|.$$

This proves at the same time the stated isometry property and the fact that the cylindrical process $(\int X\, d\tilde{M})$ can be identified with a square integrable $\mathbf{G}$-valued martingale (see Section 15.5).

16.6 Comments and a Counterexample

We have mentioned that $\tilde{\Lambda}(\tilde{M}; \mathbf{B}'; \mathbf{K}')$ is not complete. We build a counterexample involving in fact the real brownian motion for $\tilde{M}$ and a Hilbert space $\mathbf{H}$ for $\mathbf{K}$. The noncompleteness clearly depends neither on the fact that $\tilde{M}$ is cylindrical nor on the fact that the spaces $\mathbf{B}$ and $\mathbf{K}$ are non-Hilbert. It results from the "loose" character of the norm defining the topology of $\tilde{\Lambda}(\tilde{M}; \mathbf{B}'; \mathbf{K}')$. This norm is in fact the norm of pointwise convergence on a space of bounded linear operators. This is therefore not surprising. Here now is the example.

M is the real brownian motion on $[0, 1]$. Replacing if necessary the space Ω by $\Omega \times [0, 1]$, the σ-algebras $\mathcal{F}_t$ by $\mathcal{F}_t \otimes \mathcal{B}_{[0,1]}$, and P by $P \otimes l$ (where l is the Lebesgue measure), we may assume that it is possible to find a sequence I_n of events such that $\limsup_n I_n = \Omega$, $I_n \in \mathcal{F}_t$, for all t and $P(I_n) \leqslant 1/n$. We set $g_k = k^{1/3} 1_{I_k}$. We consider a separable Hilbert space $\mathbf{H}$ with an orthonormal basis (e_k) and define the following sequence of $\mathbf{H}$-valued processes:

$$\langle X_t^n, e_k \rangle_{\mathbf{H}} = \begin{cases} g_k & \text{if } n \geqslant k, \\ 0 & \text{if } n < k. \end{cases}$$

The reader will easily check that

$$N(X^{n+p} - X^n) \leqslant \int_0^1 |g_{n+1}|^2\, dP \leqslant n^{-1/2}.$$

The sequence (X^n) is therefore a Cauchy sequence for the seminorm N, and it is easy to see that the existence of a limit X in $\tilde{\Lambda}(\tilde{M}; \mathbf{R}; \mathbf{H})$ would imply $\langle X, e_k \rangle = \lim_n \langle X^n, e_k \rangle$ in $L^2(\Omega', \mathcal{P}, P \times l)$, $X(\omega, t)$ being a linear operator from $\mathbf{R}$ into $\mathbf{H}$, i.e., an element of $\mathbf{H}$. Therefore $\langle X_t, e_k \rangle = g_k$ a.s. for all t. But this would imply $\limsup_k \langle X_t, e_k \rangle = +\infty$ a.s. for all t, which is contradictory to $X(\omega, t) \in \mathbf{H}$.

HISTORICAL NOTES

Section 14. In dealing with stochastic linear partial differential equations, Curtain and Falb have considered (see [CuF]) stochastic integrals of operator-valued processes, which are not strongly predictable. It was then remarked by Pistone that this space of processes was not complete for the natural topology leading to their integration. The integration theory presented here, in which we consider processes whose values are unbounded linear operators, is extracted from the paper [MPi-2]. Besides the field of stochastic partial differential equations (see, for example, [Par-2]), this integral was used by Ouvrard (see [Ouv-1]) to obtain a representation theorem for Hilbert-valued martingales. The representation theorem of Section 14.8 is a directly proved particular version of the main theorem in [Ouv-1].

Sections 15 and 16. These two sections contain unpublished results by the authors on cylindrical martingales and integration with respect to them. For quite a while the stochastic integral with respect to cylindrical brownian motion has been introduced into the study of stochastic distributed systems (see, for example, [Ben], [Par-1]). The notion of the radonifying integral, which is extended here in Section 16.5, was first introduced by Gaveau [Gav].

BIBLIOGRAPHY

[All] M. F. Allain, Sur quelques types d'approximation des solutions d'équations différentielles stochastiques. Thèse de 3ème cycle, Université de Rennes, 1974.

[BaC] A. Badrikian and S. Chevet, *Mesures Cylindriques. Espaces de Wiener et Fonctions Aléatoires Gaussiennes*. Lecture Notes in Mathematics No. 379. Springer-Verlag, Berlin and New York, 1974.

[Bad] A. Badrikian, *Séminaire sur les Fonctions Aléatoires Linéaires et les Mesures Cylindriques*. Lecture Notes in Mathematics, No. 139. Springer-Verlag, Berlin and New York, 1970.

[Bar] R. G. Bartle, A general bilinear vector integral, *Studia Math.* **15**, 337–352 (1956).

[BDS] R. G. Bartle, N. Dunford, J. Schwarz, Weak compactness and vector measures, *Canad. J. Math.* **7**, 289–305 (1955).

[Bau] H. Bauer, *Probability Theory and Elements of Measure Theory.* Holt, New York, 1972.

[Ben] A. Bensoussan, *Filtrage Optimal des Systèmes Linéaires*. Dunod, Paris, 1971.

[BeT-1] A. Bensoussan and R. Temam, Equations aux dérivées partielles stochastiques non linéaires, *Israel J. Math.* **11**, 95–129 (1972).

[BeT-2] A. Bensoussan and R. Temam, Equations stochastiques du type de Navier–Stokes, *J. Funct. Anal.* **13**, 195–222 (1973).

[Bil] P. Billingsley, *Convergence of Probability Measures.* Wiley, New York, 1968.

[BlG] R. M. Blumenthal and R. K. Getoor, *Markov Processes and Potential Theory.* Academic Press, New York, 1968.

[Bou-1] N. Bourbaki, *Integration*, Chaps. I–VI. Hermann, Paris, 1965.

[Bou-2] N. Bourbaki, *Espaces Vectoriels Topologiques.* Hermann, Paris, 1967.

[Bur] D. L. Burkholder, Martingale transforms, *Ann. Math. Statist.* **37**, 1495–1505 (1966).

[BDG] D. Burkholder, B. Davis, and R. Gundy, Integral inequalities for convex functions of operators and martingales, *Proc. 6th Berkeley Symp.* **2**, 223–240 (1972).

[Chac] R. V. Chacon, A stopped proof of convergence, *Advances in Math.* **14**, 365–368 (1974).

[Chat] S. D. Chatterji, Martingales of Banach valued random variables, *Bull. Amer. Math. Soc.* **66**, 129–139 (1970).

[Che] S. Chevet, S. A. Chobanian, W. Linde, and V. I. Tarieladze, Caractérisation de certaines classes d'espaces de Banach par des mesures gaussiennes, *C. R. Acad. Sci. Paris Ser. A* **285**, (1977).

[Cou] Ph. Courrèges, Intégrales stochastiques et martingales de carré intégrable, Séminaire Brelot-Choquet-Deny, 7ème année (1962–1963), Exposé No. 7.

[Cur] R. F. Curtain, Stochastic evolution equations with general white noise disturbance, *J. Math. Anal. Appl.* **60**, 570–595 (1977).

[CuF] R. F. Curtain and P. L. Falb, Stochastic differential equations in Hilbert spaces, *J. Differential Equations* **10**, 412–430 (1971).

[Dal] Yu. L. Daletskii, Infinite dimensional elliptic operators and parabolic equations connected with them, *Russian Math. Surveys* **22**, 1–53 (1967).

[Dav] B. Davis, On the integrability of the martingale square function, *Israel J. Math.* **8**, 187–190 (1970).

[Daw-1] D. A. Dawson, Stochastic evolution equations, *Math. Biosci.* **15**, 287–316 (1972).

[Daw-2] D. A. Dawson, Stochastic evolution equations and related measure processes, *J. Multivariate Anal.* **5**, 1–52 (1975).

[Del-1] C. Dellacherie, *Capacités et Processus Stochastiques*. Ergebn. der Math., Vol. 67. Springer-Verlag, Berlin and New York, 1972.

[Del-2] C. Dellacherie, *Intégrales Stochastiques par Rapport aux Processus de Wiener et de Poisson.* Séminaire de Probabilités VIII, Lecture Notes in Mathematics, No. 381. Springer-Verlag, Berlin and New York, 1974.

[Del-3] C. Dellacherie, Contribution to the International Meeting of Mathematicians, Helsinki, 1978.

[DeM] C. Dellacherie and P. A. Meyer, *Probabilités et Potentiel*, Chaps. I–IV. Hermann, Paris, 1975 (édition refondue).

[Din] N. Dinculeanu, *Vector Measures*. Deutscher Verlag der Wisneuschaften, Berlin, 1966.

[Dol-1] C. Doléans-Dade, *Une Martingale Uniformément Intégrable mais non Localement de Carré Intégrable*. Séminaire de Probabilités V, Lecture Notes in Mathematics, No. 191. Springer-Verlag, Berlin and New York, 1971.

[Dol-2] C. Doléans-Dade, Intégrales stochastiques dépendant d'un paramètre, *Bull. Inst. Statist. Univ. Paris* **16**, 23–34 (1967).

[Dol-3] C. Doléans-Dade, Existence du processus croissant naturel associé à un potentiel de classe (D), *Z. Warhsch. Verw. Gebiete* **9**, 309–314 (1968).

[Dol-4] C. Doléans-Dade, On the existence and unicity of solutions of stochastic integral equations, *Z. Wahrsch. Verw. Gebiete* **36**, 93–101 (1976).

[DoM-1] C. Doléans-Dade and P. A. Meyer, *Intégrales Stochastiques par Rapport aux Martingales Locales.* Séminaire de Probabilités IV, Lecture Notes in Mathematics, No. 124. Springer-Verlag, Berlin and New York, 1970.

[DoM-2] C. Doléans-Dade and P. A. Meyer, *Equations Différentielles Stochastiques.* Séminaire de Probabilités XI, Lecture Notes in Mathematics, No. 581. Springer-Verlag, Berlin and New York, 1977.

[Doo-1] J. L. Doob, *Stochastic Processes.* Wiley, New York, 1953.

[Doo-2] J. L. Doob, Notes on martingale theory, *Proc 4th Berkeley Symp.* **2**, 95–102 (1960).

[Dre] L. Drewnowski, Topological rings of sets. Continuous set functions. Integration, *Bull. Acad. Polon. Sci.* **22**, Nos. I, II, III, 269–286, 439–445 (1972).

[DuS] N. Dunford and J. T. Schwarz, *Linear Operators*, Part 1. Wiley, New York, 1957.

[Eme-1] M. Emery, Perturbation d'équations différentielles stochastiques; intégrales multiplicatives, *C.R. Acad. Sci. Paris Ser. A*, **285**, 1977.

[Eme-2] M. Emery, Stabilité des solutions des équations différentielles stochastiques. Application aux intégrales multiplicatives stochastiques, *Z. Wahrsch. Verw. Gebiete* **41**, 241–262 (1978).

[Eme-3] M. Emery, *Une Topologie sur l'Espace des Semi-Martingales*. Séminaire de Probabilités. Springer-Verlag (to appear in Lecture Notes in Mathematics).

[Fef] Ch. Fefferman, Characterizations of bounded mean oscillation, *Bull. Amer. Math. Soc.* **77**, 587–588 (1971).

[Fis] D. L. Fisk, Quasi-martingales, *Trans. Amer. Math. Soc.* **120**, 369–389 (1965).

[Fle] W. Fleming, *Distributed Parameter Stochastic Systems in Population Biology. Internat. Symp. IRIA, 1974*, Lecture Notes in Economics and Mathematical Systems, No. 107. Springer-Verlag, Berlin and New York, 1975.

[Föl] H. Föllmer, The Exit Measure of a Super-Martingale, *Z. Wahrsch. Verw. Gebiete* **21**, 154–166 (1972).

[Gal] L. I. Galtchouk, Structure de certaines martingales, *Proc. Séminaire Processus Aléatoires, Drusnminskai, 1974* **1**, 7–32 (in russian).

[Gar] A. M. Garsia, The Burgess Davis inequalities via Fefferman's inequality, *Ark. Mat.* **11**, 229–237 (1973).

[Gav] B. Gaveau, Intégrale stochastique radonifiante, *C. R. Acad. Sci. Paris Ser. A* **276**, 617–620 (1973).

[GeW] I. M. Gelfand and N. Y. Wilenkin, *Generalized Functions*, Applications of Harmonic Analysis, Vol. 4. Academic Press, New York, 1964.

[Gir-1] I. V. Girsanov, On transforming a certain class of stochastic processes by absolutely continuous substitution of measures, *Theory Probab. Appl.* **5**, 314–330 (1960).

[Gir-2] I. V. Girsanov, An example of non uniqueness of the solution of K. Ito's stochastic integral equations, *Theory Probab. Appl.* **7**, 336–342 (1962).

[GiS-1] I. I. Gikhman and A. V. Skorohod, *Stochastic Differential Equations.* Springer-Verlag, Berlin and New York, 1972.

[GiS-2] I. I. Gikhman and A. V. Skorohod, *The Theory of Stochastic Processes-I.* Springer-Verlag, Berlin and New York, 1974.

[Glo] P. Y. Glorennec, Approximation d'équations différentielles stochastiques, Thèse de 3ème cycle, Université de Rennes, 1977.

[GlP] P. Y. Glorennec and J. Pellaumail, Théorème de Riesz pour des processus réels, *Ann. Inst. H. Poincaré* **10**, No. 3, 355–367 (1974).

[GrP] B. Gravereaux and J. Pellaumail, Formule de Ito pour des processus à valeurs dans des espaces de Banach, *Ann. Inst. H. Poincaré*, **10** No. 4, 339–422 (1974).

[Hel] L. L. Helms, Mean convergence of martingales, *Trans. Amer. Math. Soc.* **87**, 433–366 (1958).

[HeJ] L. L. Helms and G. Johnson, Class D super-martingales, *Bull. Amer. Math. Soc.* **69**, 59–62 (1963).

[Iot-1] A. Ionescu-Tulcea and C. Ionescu-Tulcea, Abstract ergodic theorems, *Trans. Amer. Math. Soc.* **107**, 107–125 (1963).

[Iot-2] A. Ionescu-Tulcea and C. Ionescu-Tulcea, *Topics in the Theory of Lifting.* Springer-Verlag, Berlin and New York, 1969.

[Ito-1] K. Ito, Stochastic integral, *Proc. Imp. Acad. Tokyo* **20**, 519–524 (1944).

[Ito-2] K. Ito, On stochastic integral equation, *Proc. Japan Acad.* **22**, 32–35 (1946).

[Ito-3] K. Ito, On a formula concerning stochastic differentials, *Nagoya Math. J.* **3**, 55–65 (1951).

[Ito-4] K. Ito, *Lectures on Stochastic Processes.* Tata Institute for Fundamental Research, Bombay, 1961.

[Ito-5] K. Ito, On stochastic differential equations, *Mem. Amer. Math. Soc.*, No. 4 (1961).

[ItM] K. Ito and H. P. Mac Kean, *Diffusion Processes and Their Sample Paths.* Springer-Verlag, Berlin and New York, 1974.

[ItW] K. Ito and S. Watanabe, Transformation of Markov processes by additive functionals, *Ann. Inst. Fourier (Grenoble)* **15**, 13–30 (1965).

[Jac-1] J. Jacod, Sous-espaces stables de martingales, *Z. Wahrsch. Verw. Gebiete* **44**, 103–115 (1978).

[Jac-2] J. Jacod, *Calcul Stochastique et Problémes de Martingales.* Lecture Notes in Mathematics. N.714. Springer-Verlag, Berlin and New York, 1979.

[JaM] J. Jacod and J. Memin, Caractéristiques locales et conditions de continuité absolue pour les semi-martingales, *Z. Wahrsch. Verw. Gebiete* **35**, 1–37 (1976).

[Kas] B. S. Kasin, *Mat. Zametki* **14**, 645–654 (1973).

[Kaz] N. Kazamaki, *Note on a Stochastic Integral Equation*, Lecture Notes Séminaire de Probabilités VI, Lecture Notes in Mathematics, No. 258. Springer-Verlag, Berlin and New York, 1972.

[Kun] H. Kunita, Stochastic integrals based on martingales taking values in Hilbert spaces, *Nagoya Math. J* **38**, 41–52 (1970).

[KuW] H. Kunita and S. Watanabe, On square integrable martingales, *Nagoya Math. J.* **30**, 209–245 (1967).

[Kus] A. V. Kussmaul, Stochastic Integration and Generalized Martingales. Pitman, London, 1977.

[Lep-1] D. Lepingle, La variation d'ordre p des semi-martingales, *Z. Wahrsch. Verw. Gebiete* **36**, 295–316 (1976).

[Lep-2] D. Lepingle, *Une Inégalité de Martingales.* Séminaire de Probabilités XII, Lecture Notes in Mathematics, No. 649. Springer-Verlag, Berlin and New York, 1978.

[LeO] D. Lepingle and J. Y. Ouvrard, Martingales browniennes hilbertiennes, *C. R. Acad. Sci. Paris* **276** 1225 (1973).

[MaS-1] V. Mandrekar and H. Salehi, The square integrability of operator-valued functions with respect to a non-negative operator-valued measure and the Kolmogorov-isomorphism theorem, *Indiana Univ. Math. J.* **20**, No. 6 (1970).

[MaS-2] V. Mandrekar and H. Salehi, Operator-valued wide sense Markov processes and solutions of infinite dimensional linear differential systems driven by white noise, *Math. Systems Theory* **4**, No. 4 340–356 (1970).

[MaO] W. Matuszewska and W. Orlicz, A note on modular spaces IX, *Bull. Acad. Polon. Sci. Ser. Sci. Math. Astronom. Phys.* **16**, No. 10 (1968).

[Mau] B. Maurey, Theoèmes de factorisation, *Astérique* **11**, 1974.

[McK] H. P. McKean, Jr., *Stochastic Integrals.* Academic Press, New York, 1969.

[McS-1] E. J. McShane, *Stochastic Calculus and Stochastic Models.* Academic Press, New York, 1974.

[McS-2] E. J. McShane, Stochastic differential equations, *J. Multivariate Anal.* **5**, 121–177 (1975).

[Mem] J. Memin, *Décompositions Multiplicatives de Semi-Martingales Exponentielles et Applications.* Séminaire de Probabilités XII, Lecture Notes, No. 649. Springer-Verlag, Berlin and New York, 1978.

[Met-1] M. Metivier, Stochastic integrals and vector-valued measures. In *Vector and Operator-Valued Measures and Applications* (D. H. Tucker and H. B. Maynard, eds.). Academic Press, New York, 1973.

[Met-2] M. Metivier, Mesure stochastique engendrée par une martingale à valeurs hilbertiennes, *C.R. Acad. Sci. Paris Ser. A* **276**, 939–941 (1973).

[Met-3] M. Metivier, Intégrale stochastique par rapport à des martingales hilbertiennes, *C.R. Acad. Sci. Paris Sér. A* **277**, 1009–1011 (1973).

[Met-4] M. Metivier, Mesure stochastique locale associée à une martingale locale à valeurs dans un espace de Hilbert, *C.R. Acad. Sci. Paris Ser. A* **273** 908–911 (1973).

[Met-5] M. Metivier, Intégrale stochastique par rapport à des processus à valeurs dans un espace de Banach réflexif, *Teor. Verojatnost. i Primenen.* **19** 577–606 (1974).

[Met-6] M. Metivier, *Reelle und vektorwertige Quasi-Martingale und die Theorie der stochastichen Integration.* Lecture Notes in Mathematics, No. 607. Springer-Verlag, Berlin and New York, 1977.

[Met-7] M. Metivier, Martingales à valeurs vectorielles. Applications à la dérivation des mesures vectorielles, *Ann. Inst. Fourier (Grenoble)* **17**, No. 2, 175–208 (1967).

[MeP-1] M. Metivier and J. Pellaumail, On Doléans–Föllmer's measure for quasi-martingales, *J. Math.* **77**, 491–504 (1975).

[MeP-2] M. Metivier and J. Pellaumail, Notions de base sur l'intégrale stochastique, Rapport I.R.I.S.A., No. 61, Rennes, 1976.

[MeP-3] M. Metivier and J. Pellaumail, Cylindrical stochastic integral, Séminaire de l' Université de Rennes, 1976.

[MeP-4] M. Metivier and J. Pellaumail, Mesures stochastiques à valeurs dans des espaces L_0, *Z. Wahrsch. Verw. Gebiete* **40**, 101–114 (1977).

[MeP-5] M. Metivier and J. Pellaumail, Une formule de majoration pour martingales, *C.R. Acad. Sci. Paris Ser. A* **285**, 685–688 (1977).

[MeP-6] M. Metivier and J. Pellaumail, On a stopped Doob's inequality and general stochastic equations, *Ann. Probab.* **7**, 1979.

[MeP-7] M. Metivier and J. Pellaumail, Sur une équation stochastique assez générale, *C.R. Acad. Sci. Paris Ser. A* **285**, 921–923 (1977).

[MPi-1] M. Metivier and G. Pistone, Sur une équation d'évolution stochastique, *Bull. Soc. Math. France* **104**, 65–85 (1976).

[MPi-2] M. Metivier and G. Pistone, Une formule d'isométrie pour l'intégrale stochastique hilbertienne et équations d'évolution linéaires stochastiques, *Z. Wahrsch. Verw. Gebiete* **33**, 1–18 (1975).

[Mey-1] P. A. Meyer, *Probabilités et Potentiel.* Hermann, Paris, 1966.

[Mey-2] P. A. Meyer, *Intégrales Stochastiques-I*, Séminaire de Probabilités I, Lecture Notes in Mathematics, No. 39. Springer-Verlag, Berlin and New York, 1967.

[Mey-3] P. A. Meyer, *Intégrales Stochastiques-II*, Séminaire de Probabilités I, Lecture Notes in Mathematics, No. 39. Springer-Verlag, Berlin and New York, 1967.

[Mey-4] P. A. Meyer, *Un Cours sur les Intégrales Stochastiques*. Séminaire de Probabilités X, Lecture Notes in Mathematics, No. 511. Springer-Verlag, Berlin and New York, 1976.

[Mey-5] P. A. Meyer, *Inégalités de Norme pour les Intégrales Stochastiques*, Séminaire de Probabilités XII, Lecture Notes in Mathematics, No. 649. Springer-Verlag, Berlin and New York, 1978.

[Mey-6] P. A. Meyer, A decomposition theorem for supermartingales, *Illinois J. Math.* **6**, 193–205 (1962).

[Mey-7] P. A. Meyer, Decomposition for supermartingales: the uniqueness theorem, *Illinois J. Math.* **7**, 1–17 (1963).

[Mey-8] P. A. Meyer, *Note sur les Intégrales Stochastiques. Intégrales Hilbertiennes*, Séminaire de Probabilités XI, Lecture Notes in Mathematics, No. 581. Springer-Verlag, Berlin and New York, 1977.

[Mil] P. W. Millar, Martingale integrals, *Trans. Amer. Math. Soc.* **133**, No. 1, 145–146 (1968).

[Nel] E. Nelson, *Dynamical Theory of Brownian Motion*, Math. Notes. Princeton Univ. Press, Princeton, New Jersey, 1967.

[Nev-1] J. Neveu, Relation entre la théorie des martingales et la théorie ergodique, *Colloque Internat. de Théorie du Potentiel, Paris, June 1974*. Editions du C.N.R.S..

[Nev-2] J. Neveu, Intégrales stochastiques et applications, Cours de 3ème cycle, Université de Paris VI, 1971–1972.

[Nev-3] J. Neveu, *Discrete-Parameter Martingales*. North-Holland, Amsterdam, 1975.

[Nev-4] J. Neveu, *Bases Mathématiques du Calcul des Probabilités*. Masson, Paris, 1964.

[Nik] E. M. Nikishin, Resonance theorems and superlinear operators. Transl. of *Uspehi Mat. Nauk.* **XXV-6** (1970).

[Ore] S. Orey, *F*-processes, *Proc. 5th Berkeley Symp. Mathematical Statistics and Probability-II* **1**, 301–314 (1965). Univ. of California Press, Berkeley.

[Ouv-1] J. Y. Ouvrard, Représentation de martingales vectorielles de carré intégrable, *Z. Wahrsch. Verw. Gebiete* **33**, 195–208 (1975).

[Ouv-2] J. Y. Ouvrard, Martingales locales et théorèmes de Girsanov dans les espaces de Hilbert réels séparables, *Ann. Inst. H. Poincaré, Sect. B* **9**, No. 4 351–368 (1973).

[Par-1] E. Pardoux, Sur des équations aux dérivées partielles stochastiques monotones, *C.R. Acad. Sci. Paris, Ser. A* **275**, 101–103 (1972).

[Par-2] E. Pardoux, Equations aux dérivées partielles stochastiques non linéaires monotones. Etude de solutions fortes de type Ito, Thèse d'Etat, Université de Paris-Sud/Orsay, 1975.

[Par] K. R. Parthasarathy, *Probability Measures on Metric Spaces.* Academic Press, New York, 1967.

[Pel-1] J. Pellaumail, Intégrale de Daniell à valeurs dans un groupe, *Rev. Roumaine Math. Pures Appl.* **16**, No. 8, 1227–1236 (1971).

[Pel-2] J. Pellaumail, Sur la décomposition de Doob–Meyer d'une quasi-martingale, *C.R. Acad. Sci. Paris Ser. A* **274** 1563–1565 (1972).

[Pel-3] J. Pellaumail, Sur l'intégrale stochastique et la décomposition de Doob–Meyer, *Astérique*, No. 9, Soc. Math. France, 1973.

[Pel-4] J. Pellaumail, On the use of group-valued measures in stochastic processes. In *Symposia Mathematica*, Vol. 21. Academic Press, New York and London, 1977.

[Pet] B. J. Pettis, On integration in vector spaces, *Trans. Amer. Math. Soc.* **44**, 277–304 (1938).

[Pis] G. Pisier, Séminaire Maurey-Schwartz, Exposé No. 6, Ecole Polytechnique, 1973–1974.

[Pop] Z. R. Pop-Stojanovic, Decomposition of Banach-valued quasi-martingales, *Math. Systems Theory* **5**, No. 4, 344–348 (1971).

[Pri] P. Priouret, *Processus de Diffusion et Equations Différentielles Stochastiques*, Ecole d'Eté de Probabilités de Saint-Flour, Lecture Notes in Mathematics, No. 390. Springer-Verlag, Berlin and New York, 1974.

[Pro-1] Ph. E. Protter, On the existence, uniqueness, convergence and explosions of solutions of systems of stochastic integral equations, *Ann. Probab.* **5**, No. 2, 243–261 (1977).

[Pro-2] Ph. E. Protter, Right-continuous solutions of systems of stochastic integral equations, *J. Multivariate Anal.* **7**, 204–214 (1977).

[Pro-3] Ph. E. Protter, H^p-stability of solutions of stochastic differential equations, preprint.

[Pro-4] Ph. E. Protter. A comparison of stochastic integrals. *Ann. Prob.* **7** No. 2, pp. 276–289 (1979).

[Rao-1] K. M. Rao, On decomposition theorems of Meyer, *Math. Scand.* **24**, 66–78 (1969).

[Rao-2] K. M. Rao, Quasi-martingales, *Math. Scand.* **24**, 79–92 (1969).

[Rud] W. Rudin, *Functional Analysis*. McGraw-Hill, New York, 1973.

[Sca] F. Scalora, Abstract martingale convergence theorem, *Pacific J. Math.* **11**, No. 1, 347–374 (1961).

[Sch] L. Schwartz, *Random Measures on Arbitrary Topological Spaces*. Tata Institute, Bombay.

[Sio] M. Sion, Outer measures with values in a topological group, *Proc. London Math. Soc.* **19**, No. 3, 89–106 (1969).

[Sko] A. V. Skorohod, *Studies in the Theory of Random Processes*. Addison-Wesley, Reading, Massachusetts, 1965.

[StV] K. Strook and S. Varadhan, Diffusion processes with continuous coefficients-I, *Comm. Pure Appl. Math.* **12** (1969).

[SuE] L. Sucheston and G. A. Edgar, A class of asymptotic martingales, *Z. Wahrsch. Verw. Gebiete* **36**, 85–92 (1976).

[Tre] F. Treves, *Topological Vector Spaces. Distributions and Kernels*. Academic Press, New York, 1967.

[Tur] P. Turpin, Suites sommables dans certains espaces de fonctions mesurables, *C.R. Acad. Sci. Paris Ser. A* **280**, 349–352 (1975).

[VSW] Van Schuppen and N. E. Wong, Transformation of local martingales under a change of law, *Ann. Probab.* **2**, 879–888 (1974).

[Vio-1] M. Viot, Solution en loi d'une équation aux dérivées partielles stochastique non linéaire: méthode de compacité, *C.R. Acad. Sci. Paris Ser. A* **278**, 1185–1188 (1974).

[Vio-2] M. Viot, Equations aux dérivées partielles stochastiques: formulation faible, Thèse d'Etat, Université de Paris VI, 1976.

[Wie] N. Wiener, Differential space, *J. Math. Phys. Math. Inst. Tech.* **2**, 131–174 (1923).

[WoZ-1] E. Wong and M. Zakai, On the convergence of ordinary integrals to stochastic integrals, *Ann. Math. Statist.* **36** (1965).

[WoZ-2] E. Wong and M. Zakai, Martingales and stochastic integrals for processes with a multi-dimensional parameter, *Z. Wahrsch. Verw. Gebiete* **29**, 109–122 (1974).

[YaW] Yamada and S. Watanabe, On the uniqueness of solutions of stochastic differential equations, *J. Math. Kyoto Univ.* **2** No. 1 (1971).

[Yor-1] M. Yor, Sur les intégrales stochastiques à valeurs dans un Banach, *C.R. Acad. Sci. Paris Ser. A* **277**, 467–469 (1973).

[Yor-2] M. Yor, *Sous-Espaces Denses dans L^1 ou H^1 et Représentation des Martingales*, Séminaire de Probabilités XII, Lecture Notes in Mathematics, No. 649. Springer-Verlag, Berlin and New York, 1978.

[Yor-3] M. Yor, Remarques sur les normes H^p de (semi)martingales, *C.R. Acad. Sci. Paris Ser. A* **287**, 461–464 (1978).

INDEX